A REMARKABLE JOURNEY

A REMARKABLE JOURNEY

THE STORY OF EVOLUTION

R. PAUL THOMPSON

REAKTION BOOKS

In memory of
Iris Monica Schmidt (1961–2011)
Who savoured deeply all the magic that life offers
and
John Bokhout (1989–2011)
Who loved much, laughed often and whose journey
had only just begun

Published by Reaktion Books Ltd
Unit 32, Waterside
44–48 Wharf Road
London N1 7UX, UK
www.reaktionbooks.co.uk

First published 2015
Copyright © R. Paul Thompson 2015

Printed and bound in Great Britain
by TJ International, Padstow, Cornwall

A catalogue record for this book is available from the British Library

ISBN 978 1 78023 446 5

Contents

Introduction

The idea that life evolved is very old. The philosopher Anaximander (c. 610–546 BCE), for example, proposed in *On Nature* that organisms evolved and that life originated in the oceans. Empedocles (c. 490–430 BCE) is reported by Aristotle in his *Physics* to have held that:

> Wherever then all the parts came about just what they would have been if they had come to be for an end, such things survived, being organized spontaneously in a fitting way; whereas those which grew otherwise perished and continue to perish, as Empedocles says his 'man-faced ox-progeny' did.

This view, which has some distant resemblance to natural selection, was hastily rejected by Aristotle: 'Such are the arguments (and others of the kind) which may cause difficulty on this point. Yet it is impossible that this should be the true view.'

After the rise of Christianity evolution fared badly in European countries, remaining dormant until after the Renaissance. It was not seriously considered until the end of the eighteenth century, when a few prominent individuals, such as Erasmus Darwin (1731–1802) – Charles Darwin's grandfather – expressed evolutionary views. For the most part, these views were inchoate and little compelling evidence was provided for them. The important developments began to occur in the nineteenth century and in these Charles Darwin (1809–1882) was the pivotal character.

Europe had begun to change rapidly in numerous ways from the middle of the eighteenth century. The Industrial Revolution transformed almost every aspect of life, from increased urbanization to the mass production of goods, such as clothing and household items. Steam engines resulted in railways replacing canals for internal transportation; steamships replaced wind-powered ships, resulting in faster and more predictable ocean travel. Although working conditions for labourers were, by today's standards, unacceptable (long hours, poor health and safety, and so on) more people had paid employment. Wealth increased, though it remained concentrated in the upper class. Nonetheless, an incipient middle class began to emerge.

Although colonialization proper had begun in 1492 when Christopher Columbus set sail for the Americas, with only minor occurrences before that time, it escalated and broadened throughout the eighteenth and nineteenth centuries. By the nineteenth century, Britain could claim that its empire was one on which the sun never sets. It extended from Britain around the globe, encompassing parts of Africa, India, China, Australia, New Zealand and North America. There was no time of day at which some part of the empire was not in daylight. The increase in resources and in the range of cuisines and foods in Britain was phenomenal. There were opponents to colonization, but supporters and opponents alike believed in the cause of 'civilizing' those in the colonies; naysayers largely focused on the abuses taking place. Although the commitment to Christianity remained firm, and for many motivated their imperialism, the knowledge of other religious views opened up a spectrum of religious possibilities. It also exposed the British to different societies and power relationships. Britain, at this time, was far from a democracy; it had a parliament and held elections but, even after the Reform Act of 1832, which introduced many changes to the electoral system, only five out of every hundred adults could vote. In 1867 – eight years after the publication of *On the Origin of Species* – another reform entitled thirteen out every hundred adults to vote, and the secret ballot was introduced only in 1872.

During the nineteenth century, theological research and analysis challenged almost two millennia of Christian thought. German theologians were in the vanguard. The dating of the Four Gospels suggested that they were written long after the events they described and were certainly not written by those whose names they bear. Analysis of the first three gospels revealed that the first (Matthew) and the third (Luke) seemed to be based on the second (Mark). Matthew and Luke clearly had other material besides Mark, which was probably the earliest gospel. Slowly, a literal reading of the Bible was being eroded.

So, by the time Darwin was exploring evolution, the world was rapidly changing, socially, economically and theologically. The shifting landscape opened up a space for scholars to entertain new ideas and for the populace to be receptive to and even fascinated by them. Evolution was one of those new ideas. Robert Chambers anonymously published his pro-evolutionary treatise, *Vestiges of the Natural History of Creation*, in 1844. It was a best-seller; even Queen Victoria read it. The milieu was ideal for a brilliant scientist to articulate a compelling theory of evolution. Charles Darwin was that brilliant scientist.

Darwin spent decades developing and collecting evidence for his theory; he opened his first notebook on evolution in 1837. This culminated in the first definitive account of the mechanisms of evolution and the evidence for its occurrence; Darwin's *On the Origin of Species* (*The Origin*) consists of remarkable discoveries, observations, experiments, conceptual ingenuity, brilliance and compelling reasoning. It was published in 1859. Darwin, however, had discovered his central mechanism, 'natural selection', 21 years earlier. In his autobiography he wrote:

> In October 1838, fifteen months after I had begun my systematic inquiry, I happened to read for amusement Malthus on Population, and being prepared to appreciate the struggle for existence which everywhere goes on, from long-continued observation of the habits of animals and plants, it at once struck me that under these circumstances favourable variations would tend to be preserved, and unfavourable ones to be destroyed. The result would be the formation of a new species.

There has been endless speculation about why it took 21 years for him to publish his views. The suggested reasons include the claim that he was a cautious scholar and wanted to be absolutely sure of his views before publishing; his dislike of controversy, which he knew the theory would bring his way; and his knowledge that publication of the theory would upset his beloved and deeply religious wife, Emma. All have some plausibility; all have significant problems. What is clear is that Darwin had written a very large manuscript on the topic by 1858. Some of his close friends knew about his theory but most scientists, and of course the public, did not. The catalyst for Darwin to publish was his receipt, on 18 June 1858, of a letter from Alfred Russel Wallace (1823–1913), who at the time was in the Malay Archipelago. In it Wallace outlined the same theory of evolution by natural selection that Darwin had been working on since 1838. Wallace wanted Darwin's opinion and requested that Darwin should arrange publication if he approved of Wallace's ideas. With the help of his close friends and confidants Joseph Hooker and Charles Lyell, an arrangement was made to have Wallace's paper, together with a rushed abstract by Darwin of his own version of the theory, read at the Linnean Society on 1 July 1858. This was done without a response from Wallace, but when he did respond, he graciously agreed. Darwin then turned his efforts to writing what he called an abstract of his 'big book', which emerged as *On the Origin of Species*. The publication of *The Origin* is the important point of departure for our story, although some earlier events and contexts open this narrative.

There are many facets to the development of evolutionary theory; each embodies a fascinating tale. This book explores the theoretical facets, those that transformed the theory at various points. These transformations led to a deeper and richer theory of evolution: the one that today permeates all biology and is an unassailable pillar of the edifice of modern science. Along the way, we will glance at the social and political implications of the changes as well as the technologies it has spawned.

I will now turn to a sketch of our journey. The watershed of our intellectual journey is the publication of *The Origin*. Most of this book is about the side of this watershed that stretches to the present, but chapter One provides a sketch of some events and ideas in the century before 1859. In chapter Two, we take a detailed look at the theory as Darwin presented it; his arguments, evidence and overall structure. Although *The Origin* is rich in evidence and compelling arguments, there are crucial parts of the theory that remained mysterious. As chapter Two makes clear, the heritability of traits (characteristics of an organism) is a central tenet of Darwin's theory, but the available theories about inheritance were sketchy and lacked compelling evidence. Darwin and his fellow scientists did not doubt that traits were heritable. It seems obvious that offspring resemble parents; moreover, breeders had been selecting desirable traits in plants and animals for a long time. The problem was an adequate understanding of how traits were passed from generation to generation. Another important component of the theory is variation. Again, the existing theories regarding variation were sketchy and lacked compelling evidence. Chapter Three examines these missing elements and the various attempts to provide both a theory of heredity and a theory that explains the maintenance of variation and the origin of novel varieties.

In one of the great ironies of science, in 1865 an Austrian monk, Gregor Mendel (1822–1884), had published a theory of heredity that later became the bedrock of modern population genetics. Darwin and most of the scientific community were unaware of his paper and of his theory. There are six recognized editions of *The Origin*. The last three were published after 1865; the fourth edition was published in 1866, the fifth in 1869 and the sixth in 1872. Had Mendel's work been known, Darwin would have had his much-needed theory of heredity by the fourth edition and certainly by the fifth. Fascinating though the thought is, there does not seem to be evidence that Darwin had a copy of the journal in which Mendel published. Chapter Four sets out Mendel's experiments and the structure of his theory.

Mendel's theory did become widely known from 1900. Far from settling the question of heredity, however, it led to acrimonious and personal feuding between two groups of scientists: the Mendelians and the biometricians. Both groups accepted evolution as a fact. The Mendelians thought that Darwin's theory was wrong, and that Mendel's theory demonstrated that fact. The biometricians, with some internal differences, accepted Darwin's theory. Chapter Five examines this controversy and others that dominated the first decade of the twentieth century and much of the second decade.

Building on experimental and theoretical work from the last few years of the second decade of the twentieth century, J.B.S. Haldane, Ronald A. Fisher and Sewall Wright resolved the tensions that had arisen in the first two decades. In 1930 Ronald Fisher put a number of the pieces together in a comprehensive treatment that united Darwin's theory, Mendel's theory and biometrics. The apt title of his book is *The Genetical Theory of Natural Selection*. With its publication, the controversies of early decades were resolved and population genetics coalesced as a field of biology. Chapter Six sets out Fisher's unified theory of evolution.

From 1930 to around 1955 most of the fields of biology were brought into the evolutionary framework; the result has become known as the modern synthesis. By 1960 'Nothing in Biology Makes Sense Except in the Light of Evolution', the title of an article of 1973 in *American Biology Teacher* by the geneticist Theodosius Dobzhansky, was a truism. Chapter Seven traces the production of this synthesis.

The entire biological landscape changed in 1953. In that year James D. Watson and Francis Crick published their groundbreaking model of the structure of DNA. It was a relatively simple model but had far-reaching consequences. It revealed how DNA codes for its own reproduction, and for the construction of proteins: the building blocks of cells and their functioning. The chemical basis of heredity was now understood. Knowledge of the structure and behaviour of DNA and of proteins led within a decade to the understanding of the chemical basis of life. In time this knowledge allowed scientists to genetically modify living things, with the current greatest successes being in plant agriculture and in medicine. Chapter Eight explains the discovery and its far-reaching importance and impact.

Darwin extended his theory to the behaviour of organisms and provided much experimental work on behaviour as well as explanations of how certain behaviours had evolved. It was not until the late 1960s, however, that a thoroughgoing application to behaviour became possible. All the previous work on behaviour coalesced in Edward O. Wilson's *Sociobiology: The New Synthesis* (1975). It created a storm of controversy but paved the way for more sophisticated evolutionary explanations of behaviour. Chapter Nine describes the evolutionary mechanism leading to behaviours and their fundamental role in evolutionary theory, as well as their limits.

DNA codes for its own replication and for the construction of proteins from more basic chemicals called amino acids, as is described in chapter Nine, but in all but the simplest organisms there is a complex process through which organisms are constructed. That process is called 'development' by biologists. Sometimes for clarity it is called 'embryological development'. This is the process that begins with a single cell – a fertilized ovum (egg) – and ends with a self-sustaining, independent organism. The more we learn about this process, the more important it seems; it is at least as important as genes in producing an organism. It is now clear that specific collections of genes control the role other genes play in the development of the organism. And there are many other ways in which the developmental process contributes to the final organism. Chapter Ten documents many of the important ones.

At this point our biological journey ends. Of course, new knowledge and the ability to explain things that it makes possible will continue to accumulate. Our journey, however, takes us into the twenty-first century. My hope is that by the time a reader makes it to the end of chapter Ten, she or he will have no doubt about the validity of the theory of evolution that has descended from Darwin, about its pervasiveness in biology and its scientific importance. This should make the rejection of evolution by nearly half the citizens of the United States inexplicable. Even a cursory examination reveals that it is a commitment to an extreme form of Christianity that lies at the heart of this rejection. There are other elements, such as regionally impoverished science education, but a commitment to a literal reading of the Christian Bible is central. Chapter Eleven describes the rejection and dismantles literalism. Over time it will become more and more difficult, with any modicum of rationality, to hold to biblical literalism and to continue to close one's eyes to the compelling case for evolution, and this chapter explains why.

Let the journey begin.

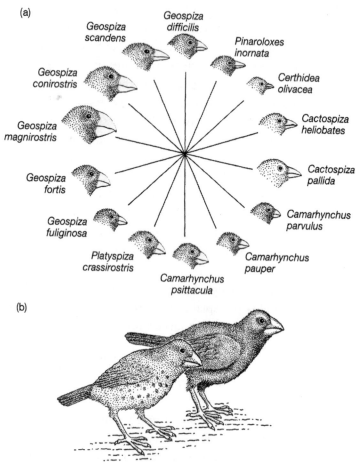

(a)

Geospiza difficilis

Geospiza scandens

Pinaroloxes inornata

Geospiza conirostris

Certhidea olivacea

Geospiza magnirostris

Cactospiza heliobates

Geospiza fortis

Cactospiza pallida

Geospiza fuliginosa

Camarhynchus parvulus

Platyspiza crassirostris

Camarhynchus pauper

Camarhynchus psittacula

(b)

1 The differing beaks of the Galápagos finches.

ONE

Prologue

It was the second week of October and oppressively hot. On two days that week, Darwin noted that the thermometer within the tent was at 93°F (33.9°C) and outside it was at 85°F (29.4°C). The temperature, however, was not his main preoccupation. His initial interest was in the geology of the islands but it soon turned to the flora and fauna, especially a curious feature of the finches. Thirteen species shared a similar beak structure, short tail and form of body and plumage. With one exception, they were all found only on these islands. The most striking feature, however, was the perfect gradation in the size of the beaks from one as large as that of a hawfinch to another as small as that of a chaffinch; there were no fewer than six species which, when ordered by beak size, showed a gradual increase (illus. 1). This increase was small between each bird in the sequence but large between the largest and the smallest. What meaning could be made of this? He mused that 'seeing this gradation and diversity of structure in one small, intimately related group of birds, one might really fancy that from an original paucity of birds in this archipelago, one species had been taken and modified for different ends.' But what could be the cause of this modification? It would be 24 years before he would share his answer with the world. In this place, however, from 15 September to 20 October 1835, the seed of curiosity and speculation about how 'from so simple a beginning endless forms most beautiful and most wonderful have been, and are, evolved' had been sown in the young Charles Darwin.

In a previous generation and half a world away from the Galápagos Islands, on English soil, another Darwin had contemplated the meaning of a remarkable discovery. In a letter to Josiah Wedgwood on 2 July 1767, Charles Darwin's grandfather, Erasmus Darwin, wrote:

> I have lately travelled two days journey into the bowels of the earth, with three most able philosophers, and have seen the Goddess of Minerals naked, as she lay in her inmost bowers, and have made such drawings and measurements of her Divinity-ship, as would much amuse. I had like to have said inform you.

He had journeyed into the Harecastle Tunnel in Staffordshire, which was being dug as part of the construction of the Grand Trunk Canal in the Pennines mountain range in mid-northern England. The excavators found fossilized plants and bones; the bones were large and old, older than an earth created by God in 4004 BCE could accommodate. Erasmus Darwin studied these fossils and, like Charles in the Galápagos 68 years later, found transmutation (evolution) of organisms an irresistible conclusion. Around 25 years passed before, in his massive two-volume *Zoonomia; or, The Laws of Organic Life* (1794–6), he published his evolutionary views. Scattered throughout the work are his views on the evidence for the fact of evolution and his hypothesized causes; 25 pages (pp. 480–505) near the end of volume I, however, is the closest he came to giving a somewhat structured and compact exposition.

By the beginning of the nineteenth century the intellectual landscape in Britain and Europe had changed substantially. The effects of the Industrial Revolution – steam engines, mechanization of production and factories – were already profound. The economic view of Adam Smith – the idea that rational self-interest in a free-market economy leads to economic equilibrium and well-being – had taken hold. The British Empire was extensive, fuelling global commerce. The members of the Lunar Society, so called because its members met each month under the full moon, had transformed science and technology, as well as economy. Its most well-known members were Erasmus Darwin, Joseph Priestley, James Watt and Josiah Wedgwood. Britain in 1837 was dramatically different from how it had been a century earlier. In that year, in that transformed society, Charles Darwin started his notebook 'on the variation of animals and plants under domestication and nature'.

Although it had not yet come to affect British society, at the beginning of the nineteenth century, in Tübingen, Germany, Friedrich Schleiermacher, David Friedrich Strauss and Ludwig Feuerbach had ushered in higher criticism in biblical studies. Higher criticism held that the Bible was a historical document that was neither literally true nor divinely inspired. It was a text to be examined using historical methods. Strauss argued that the Bible was not literally true, that Jesus was not divine and that the layers of myth and legend in the Bible made knowledge of the historical Jesus impossible. The poet Samuel Taylor Coleridge, having spent time in Germany at the turn of the century, brought higher criticism to Britain. However, it was not until George Eliot translated Strauss's *Life of Jesus* in 1846 and Feuerbach's *The Essence of Christianity* in 1853 that the influence of higher criticism took hold in Britain. By 1860 a number of Anglican theologians had adopted it.

Higher criticism called into question the literal truth of the Bible. Anomalies in the accounts of Jesus' words and deeds were explored. Matthew, Mark and Luke have Jesus cleansing the temple at the end of his ministry; John has it at the beginning. Matthew, Mark and Luke are similar in the structure of the story; John is entirely different. Mark, the oldest of the Gospels, has a less fulsome account of the Resurrection than Matthew and Luke, and the verses following 16:8 are disputed; only Matthew has the Nativity and the Sermon on the Mount. An event as important and spectacular as the raising of Lazarus is only found in John's Gospel. These anomalies posed no problem for those who saw the Bible as a human attempt to record history as they came to understand and interpret it. For literalists, however, the ad hoc explanations required were piling up and looking more and more like the manoeuvres of the desperate.

In this socially and intellectually transformed Britain, Charles Darwin pondered the causes of organic diversity. In 1842 he wrote a brief sketch of his theory, which he expanded significantly in the summer of 1844. Coincidentally in that year, evolutionary thinking was rocketed to public prominence by the publication of *Vestiges of the Natural History of Creation*. Published anonymously, it was quickly attributed to the British publisher Robert Chambers, an attribution that became official in 1884; it was stridently evolutionary.

Chambers held that natural laws govern the physical and organic worlds as well as man's activities (including his moral character). Hence organic origins are governed by natural laws. Miracles are excluded. Moreover, he held that the fossil record was progressive and evolutionary, and he sketched the Great Chain of Being manifest in it: a chain that begins with lowly creations and ends with humans. Most importantly, he held that the transmutation of species occurs and is in accordance with nature's laws. For him, embryological development was a continuous process and organic lineages paralleled embryological ones; that is, the sequence of changes in the development of an embryo is the same as the sequence of changes during the evolutionary development of its species. The only difference is that the evolutionary sequence took place over a long time, while the same sequence for a developing embryo occurs very quickly. He claimed that when, by some natural process, a new stage is added to the end of the sequence of an embryo's development, a new species arises.

The book was one of the best-sellers of its day. Everyone of note read it, including Queen Victoria. It was praised, ridiculed, criticized and defended. The idea that organisms evolved was no longer being ignored. Chambers's views were bold and novel, but they were short on evidence and many were highly speculative. They made discussions of evolution lively but its acceptance was akin to an article of faith. Fifteen years later, all that changed: *On the Origin of Species* contained a wealth of evidence, compelling arguments and a credible mechanism. Moreover, its author, Charles Darwin, was one of the most respected scientists of the day.

TWO

A New Dawn: Darwin Transforms Biology

The theory put forward in *On the Origin of Species* marks the dawn of contemporary evolutionary biology. Everything before it was the prologue, a setting of the stage. In this chapter we examine Darwin's theory, the essence of which can be captured in these six claims.

1) Organisms differ from one another. That is, different organisms have different traits. (This is variation in a population.)

2) Organisms are in competition with each other and also struggle against a hostile environment. For example, obtaining adequate food and water and finding a mate (or mates) pit organisms against each other. In addition, surviving in a hostile environment poses severe challenges: predators, disease and weather, for example, are all challenges organisms face every minute of every day. (This is the struggle for existence.)

3) Some traits are more beneficial than others in outcompeting others and in meeting the environmental challenges. Success in these struggles means having offspring. Living to an old age but without offspring is *not* survival from an evolutionary perspective. (This is natural selection; in more modern terms, differential reproductive success.)

4) Offspring will have many of the traits of their parents. Hence organisms that reproduce will contribute many of their traits to the next generation. (This is heredity.)

5) Over a large number of generations, advantageous traits will come to dominate.

6) As these advantageous traits accumulate over remarkably many generations, a new kind of organism will emerge. (This is evolution.)

This is the skeleton. *The Origin* provides the flesh.

Darwin's argument for evolution by means of natural selection is carefully constructed. He was influenced by the philosophies of science of his day. We shall uncover these and will follow the methodical steps that Darwin followed in *The Origin*. The result is a robust theory that met the well-accepted standards of a scientific theory. By the end of this chapter, the theory will have been explained in detail. At that point, it will have become clear that this is a robust theory crafted by a brilliant intellect. Nonetheless, as we shall see, there are some difficulties. First, let's explore the theory.

Darwin's theory had a long gestation. Darwin opened his first notebook on the origin of species on 1 July 1837. In 1842 he wrote a 35-page abstract of his theory. In the summer of 1844 he expanded this abstract to 230 pages. Between 1844 and 1858, but mostly in the three years leading up to 1858, he expanded the manuscript considerably and by 1858 he had a very large manuscript that is now known as his 'big book on species'. *On the Origin of Species* of 1859 was an abstract of this big book, as Darwin states in the introduction: 'This abstract, which I now publish, must necessarily be imperfect.' Darwin wanted to put the word 'abstract' in the title but John Murray, his publisher, vetoed the idea on the grounds that a 490-page book (excluding the subject index) cannot be regarded as an abstract. The section of the big book that dealt with variation was published in two volumes in 1868 under the title *The Variation of Animals and Plants under Domestication*. The section on natural selection remained unpublished until 1975 (except what was used in *The Origin*), when R. C. Stauffer, using manuscript notes, published it. Anyone who has used the Darwin archives at the University of Cambridge knows the skill it takes to read Darwin's handwriting. Stauffer's task was not a simple one.

The Origin is a masterpiece; its overall structure, its arguments and its evidence are impressive. They reveal the thought processes of an intellectual genius. Darwin was greatly influenced in his philosophy of science – his understanding of causes, theories, methodology and reasoning – by John F. W. Herschel (especially *A Preliminary Discourse on the Study of Natural Philosophy* of 1831) and William Whewell (especially the *Philosophy of the Inductive Sciences* of 1840). Herschel was a mathematician turned astronomer who for a period towards the end of his career was head of the British mint. Whewell was a professor of mineralogy at the University of Cambridge. The exemplar of a scientific theory for both was Newtonian mechanics.

Isaac Newton (1643–1727) succeeded in formulating some very general regularities (laws) from which all other regularities could be derived. A regularity is a description that is always true about the way nature always works. Newton identified four very general regularities:

1) All bodies remain in uniform rectilinear (straight-line) motion or rest unless acted upon by an external unbalanced force.

2) Force equals mass multiplied by acceleration: $f = ma$.

3) For every action, there is an equal and opposite reaction.

4) There is a force of gravitational attraction between bodies equal to the product of their masses divided by the distance between them squared: $f_g = (m_1 x m_2)/(d_1 - d_2)^2$.

From these, every other regularity, in principle, can be deduced: Galileo's law of free fall ($d = \frac{1}{2}gt^2$) and the movement of billiard balls on a billiard table are examples. These are frequently referred to as the axioms of Newton's mechanics. Axioms are the most general statements about how things behave. Their only justification is that the less general statements that are deduced from them are consistent with the observed behaviour of things in the world. Theories, understood this way, integrate a large body of knowledge and as a result provide robust explanations and predictions of the behaviour of things. So schematically, a theory for both Herschel and Whewell looked like the diagram overleaf (illus. 2).

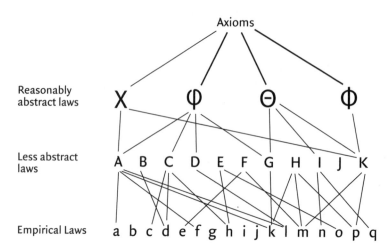

2 A schematic diagram of the deductive interconnections in a scientific theory.

Reasonably abstract laws can be deduced from the axioms. Less abstract laws can be deduced from the reasonably abstract laws and empirical laws can be deduced from less abstract laws. This is a simplification. Sometimes the steps from the axioms to the empirical laws are numerous. Nonetheless, the simplication will serve our purposes. On this view of theories, everything can be deduced from the axioms.

One place where Herschel and Whewell differ is how the axioms are discovered. Herschel was an empiricist; he believed that one starts with experience. Simple regularities (a, b, c, . . . in the diagram) are discovered from experience; if I release any object I hold above the ground, it will fall to the ground, for example. Examination of a large collection of these reveals more general statements (A, B, C, . . . in the diagram) from which groups of simple regularities can be deduced. Examination of these (A, B, C, . . .) reveals even more general statements (X, φ, Θ, . . .), all the way up to the most general statements of all – axioms.

The extreme opposite to this is rationalism that holds that one *discovers*, using reason, the axioms and then examines whether the way things in the world behave can be deduced accurately from them: that is, whether they explain the behaviour of actual things. If not, reason devises another set of axioms. Whewell held a middle position. He believed that the axioms came into focus, through reason, after a certain body of knowledge was available from observation and experimentation. Newton, for example, used reason, experiment and experience to figure out just why things behaved the way they did; he did not laboriously work his way up to the axioms in the way Herschel's view requires.

Whewell had two other elements in his philosophy that Darwin used to powerful effect. First, the power of a theory increases the more domains of empirical experience it can explain; Whewell called this the 'consilience of inductions'. This will be clearer when we see how Darwin satisfies it. Second, analogy can reveal hidden causes. The analogy is usually between some cause that we can observe and a cause that we cannot observe. Again, this will become clearer when we see how Darwin employs analogies. Darwin adopted both these elements of Whewell's philosophy of science, as well as his middle way between empiricism and rationalism.

With this view of the structure of a scientific theory in mind, Darwin tackled the most challenging features of the biological world: design and purpose. These were fundamental features of the array of living things. In Darwin's early years he accepted that design and purpose were evidence of the creative hand of the Judaeo-Christian God. God created each species as a natural kind with its own essential characteristics. Darwin slowly came to reject that view. A major influence was his reading of Charles Lyell's *Principles of Geology* (1830). Lyell presented a geological explanation that relied on two key principles. The first was actualism: the causes of geological events of the past are the same as the ones observed in action today. The second was uniformitarianism: the causes operating in the past were of the same intensity as those observed today. In effect these ruled out miracles as geological explanations. Interestingly, Lyell believed that the origin of species was supernatural. The torment that his friend Darwin's views inflicted on him on the 'species question' was profound. The other event that moved Darwin decisively away from the belief that the Christian God had created species was the death of his beloved daughter Anne Elizabeth, known as Annie, from tuberculosis in 1851 at the age of ten. She suffered for a little more than a month before her death. Darwin found that he could not believe in a God who would inflict suffering and death on such an innocent. With God removed as an explanation of design and purpose, some entirely natural explanation was required. That is what he provided.

The element of Darwin's theory that has received the most public attention is natural selection. It is undeniably a fundamental part of the theory. Nonetheless, there is a more fundamental element: species. Evolution requires that one species can transform into another. If species were specially created kinds of organisms with essential characteristics, evolution would be impossible. Natural selection might be a mechanism with the power to transform organisms but if there is an unbridgeable divide – a barrier at the species level – then it cannot produce new species. Many Christians accept that the environment can change humans (and other organisms). Humans can become lactose-tolerant or develop darker or lighter skin, for example. A human remains a human, however. There has been evolution *within* a species but no evolution *of* a species – a point to which we will return.

Darwin knew the challenge well and was keenly aware that evolution required that the barrier be removed. On 11 January 1844 he wrote to Joseph Hooker: 'At last gleams of light have come, and I am almost convinced (quite contrary to the opinion I started with) that species are not (it is like confessing a murder) immutable [unchangeable].' In *The Origin* he provides his argument for this conviction and expresses clearly the importance of this view. So significant is this issue that he begins and ends *The Origin* with mention of it; in between he produces a compelling case for such a belief. In the introduction, he wrote:

> I can entertain no doubt, after the most deliberate study and dispassionate judgement of which I am capable, that the view that most naturalists entertain, and which I formerly entertained – namely, that each species has been independently created – is erroneous. I am fully convinced that species are not immutable.

In the fourteenth and final chapter, 'Recapitulation and Conclusion', he wrote:

Although I am fully convinced of the truth of the views given in this volume under the form of an abstract, I by no means expect to convince experienced naturalists whose minds are stocked with a multitude of facts all viewed during a long course of years, from a point of view directly opposite to mine. It is so easy to hide our ignorance under such expressions as the 'plan of creation,' 'unity of design,' etc., and to think that we give an explanation when we only re-state a fact. Anyone whose disposition leads him to attach more weight to unexplained difficulties than to the explanation of a certain number of facts will certainly reject the theory. A few naturalists, endowed with much flexibility of mind, and who have already begun to doubt the immutability of species, may be influenced by this volume; but I look with confidence to the future, to young and rising naturalists, who will be able to view both sides of the question with impartiality. Whoever is led to believe that species are mutable will do good service by conscientiously expressing his conviction; for thus only can the load of prejudice by which this subject is overwhelmed be removed.

The case he assembled for the mutability of species is elegant and simple – part observation, part reasoning. The crux of the observational part is:

Hence, in determining whether a form should be ranked as a species or a variety, the opinion of naturalists having sound judgement and wide experience seems the only guide to follow. *We must, however, in many cases, decide by a majority of naturalists, for few well-marked and well-known varieties can be named which have not been ranked as species by at least some competent judges.*

That varieties of this doubtful nature are far from uncommon cannot be disputed. Compare the several floras of Great Britain, of France or of the United States, drawn up by different botanists, and see what a surprising number of forms have been ranked by one botanist as good species, and by another as mere varieties. Mr H. C. Watson, to whom I lie under deep obligation for assistance of all kinds, has marked for me 182 British plants, which are generally considered as varieties, but which have all been ranked by botanists as species; and in making this list he has omitted many trifling varieties, but which nevertheless have been ranked by some botanists as species.

Whether something is a species or a variety is frequently not obvious, even to trained naturalists. Moreover, in many cases, the best one can achieve is a majority consensus. If species are special creations with essential characteristics given to them by God, why is it so difficult in so many cases to tell whether some group of organisms is a species or a variety within a species? Darwin reasoned that the difficulty arises because species are not special creations with essential characteristics: species, sub-species and varieties are human constructs.

Finally, then, varieties have the same general characters as species, for they cannot be distinguished from species except, firstly, by the discovery of intermediate linking forms, and the occurrence of such links cannot affect the actual characters of the forms which they connect; and except, secondly, by a certain amount of difference, for two forms, if differing very little, are generally ranked as varieties, notwithstanding that intermediate linking forms have not been discovered; but the amount of difference considered necessary to give to two forms the rank of species is quite indefinite . . .

Certainly no clear line of demarcation has as yet been drawn between species and sub-species – that is, the forms which in the opinion of some naturalists come very near to, but do not quite arrive at, the rank of species; or, again, between sub-species and well-marked varieties, or between lesser varieties and individual differences. These differences blend into each other in an insensible continuous series; and a series impresses the mind with the idea of an actual passage.

Species, sub-species and varieties are human impositions, just as the named stages are in embryological development. Human development from a fertilized egg to an adult is continuous but those who study development carve it up into stages. This is scientifically useful: it aids scientific discussion and provides a vocabulary to allow parts of the series to be studied and described separately (illus. 3). But these divisions, though useful, do not change the fact that this is a *continuous* series; there are no barriers between a foetus and neonate (newborn). Similarly, even though we classify organisms as species, sub-species and varieties, there are no barriers between them.

> As from these remarks it will be seen that I look at the term species, as one arbitrarily given for the sake of convenience to a set of individuals closely resembling each other, and that it does not essentially differ from the term variety, which is given to less distinct and more fluctuating forms.

Darwin first encountered this impression of an insensible gradation on those hot days in the Galápagos Archipelago, about which he wrote in his *Voyage of the Beagle* (2nd edn): 'seeing this gradation and diversity of structure in one small, intimately related group of birds, one might really fancy that from an original paucity of birds in this archipelago, one species had been taken and modified for different ends.'

The conceptual importance of the mutability of species to the theory of evolution cannot be overemphasized, as Darwin knew well. Natural selection may be a central mechanism of organic change but without species mutability, evolution cannot occur. A classic textbook example of natural selection illustrates this well. It describes the transformation of the peppered moth (*Biston betularia*) population in Britain during the Industrial Revolution (illus. 4). In the eighteenth century and into the nineteenth, *typica*, a light-coloured

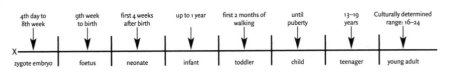

3 Human development.

variety of the moth, was the most common. In the second half of the nineteenth century a dark-coloured variety, *carbonaria*, became more numerous. By the end of the century it had become the dominant variety, with percentage counts in the high nineties in some regions. During the second half of the twentieth century *typica* again became the dominant variety. The standard explanation points to the blackening of the environment during the heavy coal-burning period of the Industrial Revolution. *Typica* was more visible to its bird predators in that environment; *carbonaria* was more camouflaged. As a result of cleaner fuels (natural gas and hydroelectricity, for example) and the anti-pollution regulations of the second half of the twentieth century, soot levels declined dramatically. *Carbonaria* was again more visible to its bird predators in that less blackened environment; *typica* was more camouflaged. Hence *typica* increased and *carbonari* declined. This is a clear instance of natural selection: a trait of an organism (colour) is being selected through the predator–prey relationship – a purely natural process. Recent research by Michael Majerus indicates that, although camouflage is a key part of the explanation, there are probably other factors at work as well, such as migration. More research will be required to determine the contribution of these other factors. Nonetheless, as Majerus clearly states, this is still an excellent example of natural selection.

Compelling though this example is, a moth remained a moth. Darwin's finches also remained finches but he considered them a different species. Darwin held, of course, that over time natural selection would change a sufficient number of the traits of a group of organisms that it would warrant being called a new species. That, however, requires that species are mutable. If there is an impenetrable barrier between species, natural selection can modify varieties within a species but never produce a new one. That is why he devoted significant attention to the nature of species and demonstrated that species are artefacts of a human system of classification. Darwin, by the end of chapter Two, has made his case about species and is ready to set out the central mechanism of change in chapter Four.

The transformation of the peppered moth illustrates natural selection. Darwin, however, did not have this example at hand. So, to uncover the conceptual underpinnings of natural selection, he

4 The transformation of the peppered moth.

provided a two-pronged approach. First, he imported into biology the struggle for existence from Thomas Malthus's economics. Malthus (1766–1834) had argued that population numbers increase geometrically while resources – the means of subsistence – grow only arithmetically. At some point the increase in population means that the resulting increase in demand for resources will outstrip the supply (illus. 5). At that point a struggle for existence will ensue. That struggle, Darwin realized instantly, is not only economic and does not apply only to human populations; it can be generalized to all organisms.

Where the two curves in the graph meet, resource demand due to increases in the population is equal to resource supply. After that point, each generation will produce more offspring than the resources

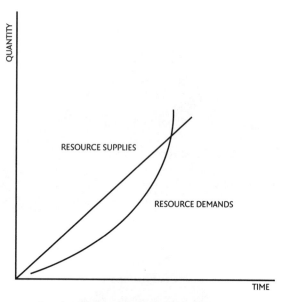

5 Supply and demand curves according to Malthus.

can support and a struggle for existence will ensue. Organisms that have traits – characteristics – that enable them to outcompete others that lack those traits will have a higher probability of leaving offspring with those same beneficial traits. This results in selection of some organisms over others. Although this is called 'natural' selection, this, as Darwin noted, is a metaphor. There is no thing, 'nature', that is selecting. It is just the dynamics of the situation: too many organisms, not enough resources, and some organisms that have advantages in the struggle to obtain them.

The second prong of his approach is an analogy. The first prong establishes that the nature of things leads to selection. The actual process of selection in nature, however, is very slow and cannot in any meaningful sense be observed. So how can we know what its effects will be? Humans have been domesticating animals for millennia; the dog was probably the first, about 16,000 years ago. Between 10,000 and 5,000 years ago goats, pigs, sheep, cows and donkeys were domesticated, in that order. Breeders selected animals with desirable traits – milk production and wool quality, for example. The same is true of plants – wheat, for example. So the power and effect of 'human' selection was well known by Darwin's era. Following

Whewell's philosophy of science, Darwin saw that a known cause of changes in organisms – human selection – was analogous to a hidden (or at least not readily observable) cause of changes in organisms, namely natural selection. If human selection produces, at an observable pace, significant changes in organisms – as we know it has – there is every reason to accept that selection in nature will have the same effect, although at a much slower pace.

There is another evidential element that the analogy of natural selection with artificial selection gives Darwin: heredity. Few doubted that offspring resemble parents, but Darwin had no account of heredity available to him. An account – indeed a theory – would be published before he finished the sixth edition of *The Origin*, but Darwin was unaware of the groundbreaking experiments and theorizing of Gregor Mendel. Darwin was able to bolster the common acceptance that characteristics were inherited by pointing to artificial selection. No matter how much humans selected beneficial traits, if they were not passed on to the next generation, breeding would be impossible. But not only was selective breeding possible and effective, it had been practised for millennia. Indeed, the techniques used had become very sophisticated. Even without a theory of heredity, Darwin could count on general acceptance of its reality.

By the end of chapter Four, Darwin had covered a lot of conceptual territory. He had established that there was much variation in nature: that is, organisms within species differ considerably from each other. He had reconceptualized species as a human carving up of an insensible gradation in nature. Hence species are mutable. Transforming one species into another is therefore possible. Natural selection is a powerful mechanism for changing the characteristics of organisms over time. The struggle for existence and the successes of artificial selection provided compelling evidence of that power. Heredity, though still something of a mystery, is real. The common acceptance of it was bolstered by selective breeding.

Although they needed to be drawn out, Darwin now had his axioms, which logically entailed the evolution of organisms. The evolution of specific organisms requires specific circumstances: organisms, traits and environments. But, like Newton, he now had the high-level generalizations. For Newton, these entailed the movement

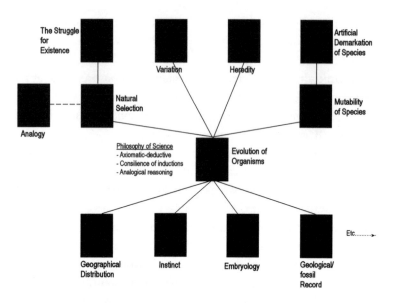

6 The axioms and structure of Darwin's theory; the top row gives the axioms.

or rest of objects. As with evolution, specific movements and behaviours depend on the specific circumstances. So Darwin's axioms and the evolution of organisms that they entail look something like that portrayed above (illus. 6).

Again drawing on Whewell's philosophy of science, particularly the consilience of inductions, Darwin knew that the more domains of natural history (biology) his theory could explain, the more robust it would be. For example, although his theory was devised to explain the observed gradations in organisms, it also explained their biogeographic distribution. The Galápagos provide an excellent example. They exhibit, as noted before, a gradation in form. The map opposite shows that their geographic distribution is also interesting (illus. 7). Cocos is further away than its location in the figure appears. It is in a square to indicate that it is an insert. The percentages indicate the number of species found only on that island: endemic species. All the species on Cocos are found only on Cocos; all the species on Indefatigable are found on at least one other island. What is clear is that the further the islands are from each other, the higher is the percentage of species found only on that island. Darwin's theory explains this kind of distribution. Finches on islands close together

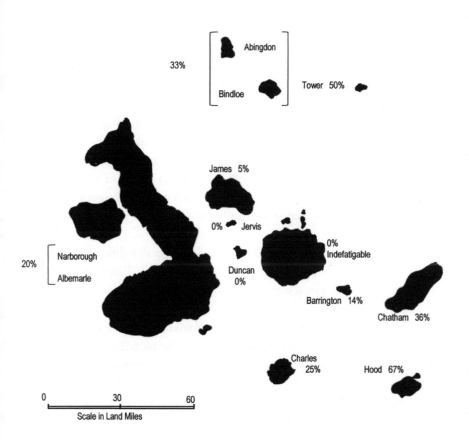

Culpepper

Wenman

75%

Cocos

100%

33%

Abingdon

Bindloe

Tower 50%

James 5%

0% Jervis

0%
Indefatigable

20%

Narborough

Albemarle

Duncan
0%

Barrington 14%

Chatham 36%

Charles
25%

Hood 67%

0   30   60
Scale in Land Miles

7 The percentage of species found only on the specific island.

will traverse the distances frequently enough to constitute a single inter-breeding population, the members of which will evolve as a single unit. As the distances increase, fewer finches will traverse the distance and interbreeding between organisms from the different islands will be less frequent, giving rise to sub-populations that evolve independently. Cocos is so far away from the main cluster of islands that no finches make the crossing – or so few that they are irrelevant from a gene-pool perspective. As always, there are factors other than distance that affect the movement of finches among the islands – such as wind patterns – but the principal factor is distance. Darwin's theory of evolution explains this kind of distribution.

Darwin's theory also explains anatomical features of organisms such as homologies – similarities in structure, as shown opposite (illus. 8). He spends chapter Twelve examining this aspect, along with embryology and rudimentary organs, concluding:

> On this same view of descent with modification, all the great facts in Morphology become intelligible, whether we look to the same pattern displayed in the homologous organs, to whatever purpose applied, of the different species of a class; or to the homologous parts constructed on the same pattern in each individual animal and plant.

According to Darwin's theory, one would expect that different species – even genera (singular genus, the next level up in classifying organisms) – would share common templates, which evolution had modified in different directions over time.

Darwin also held that his theory explains instinctive behaviour. He decided to devote a separate chapter, chapter Seven, to instinct because he understood that many would find instinct incompatible with evolution by natural selection, 'especially as so wonderful an instinct as that of the hive-bee making its cells will probably have occurred to many readers, as a difficulty sufficient to over-throw my whole theory'. Darwin, however, thought that instinctive behaviour, such as 'the instinct which leads the cuckoo to lay her eggs in other birds' nests; the slave-making instinct of certain ants; and the comb-making power of the hive-bee', could be explained by his theory. Instincts such as these, he

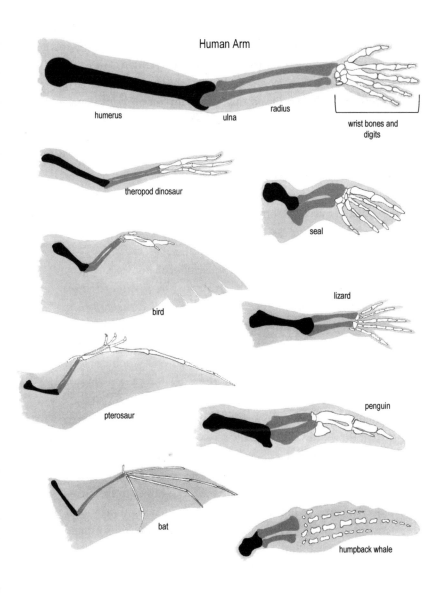

Human Arm

humerus   ulna   radius   wrist bones and digits

theropod dinosaur

seal

bird

lizard

pterosaur

penguin

bat

humpback whale

8 Homologies of limb structure.

explained, are selected because they benefit the group. Being part of a group with that instinct benefits the individual in turn. Darwin provided a wealth of experimental evidence to support this explanation of instinctive behaviour. It would be a mistake to interpret Darwin as advocating group selection along the lines of V. C. Wynne-Edwards's treatment of it in *Animal Dispersion in Relation to Social Behavior* (1962). Wynne-Edwards and others believed that the evolution of social behaviour required that natural selection acted on characteristics of groups rather than on individuals – groups were selected. More details on this are provided in chapter Nine. Darwin never uses the phrase 'group selection' in this chapter. 'Community selection' is used but his constant focus on natural selection in individual organisms is apparent in his summation of the discussion of the cell-constructing instinct in hive-bees (emphasis added):

> Thus, as I believe, the most wonderful of all known instincts, that of the hive-bee, can be explained by *natural selection having taken advantage of numerous, successive, slight modifications of simpler instincts*; natural selection having by slow degrees, more and more perfectly, led the bees to sweep equal spheres at a given distance from each other in a double layer, and to build up and excavate the wax along the planes of intersection. The bees, of course, no more knowing that they swept their spheres at one particular distance from each other, than they know what are the several angles of the hexagonal prisms and of the basal rhombic plates. The motive power of the process of natural selection having been economy of wax; that individual swarm which wasted least honey in the secretion of wax, having succeeded best, and having transmitted by inheritance its newly acquired economical instinct to new swarms, which in their turn will have had the best chance of succeeding in the struggle for existence.

For Darwin, selection acts on individuals, though sometimes an instinctive behaviour that is widespread in a group gives the members of the group, as individuals, a survival advantage over the members of other groups without that instinct. Groups are not the principal focus of selection but do contribute to the outcome for individuals. In the absence of modern population genetics, this was as far as Darwin could go. It was a significant enough distance to demonstrate that his theory could explain difficult cases of instinctive behaviour. At worst he had disarmed the objections based on instincts to his theory. He correctly thought that he had provided evidence and an account that strengthened his theory:

> On the other hand, the fact that instincts are not always absolutely perfect and are liable to mistakes; that no instinct has been produced for the exclusive good of other animals, but that each animal takes advantage of the instincts of others; that the canon in natural history, of 'natura non facit saltum' [nature does not make jumps] is applicable to instincts as well as to corporeal structure, and is plainly explicable on the foregoing views, but is otherwise inexplicable, all tend to corroborate the theory of natural selection.
>
> This theory is, also, strengthened by some few other facts in regard to instincts; as by that common case of closely allied, but certainly distinct, species, when inhabiting distant parts of the world and living under considerably different conditions of life, yet often retaining nearly the same instincts.

Darwin also held that his theory explained the fossil record. It explained what fossils are there, why there are gaps in the sequences and the temporal ordering of fossils. Darwin here drew on his extensive knowledge of, and contributions to, geology. In chapter Nine, he deals with the imperfections in the fossil record and in chapter Ten with geological succession and the record of organic life. His conclusions are:

> Passing from these difficulties, all the other great leading facts in palaeontology seem to me simply to follow on the theory of descent with modification through natural selection . . .
>
> We can understand how it is that all the forms of life, ancient and recent, make together one grand system; for all are connected by generation. We can understand, from the

continued tendency to divergence of character, why the more ancient a form is, the more it generally differs from those now living. Why ancient and extinct forms often tend to fill up gaps between existing forms, sometimes blending two groups previously classed as distinct into one; but more commonly only bringing them a little closer together . . .

If then the geological record be as imperfect as I believe it to be, and it may at least be asserted that the record cannot be proved to be much more perfect, the main objections to the theory of natural selection are greatly diminished or disappear. On the other hand, all the chief laws of palaeontology plainly proclaim, as it seems to me, that species have been produced by ordinary generation: old forms having been supplanted by new and improved forms of life, produced by the laws of variation still acting round us, and preserved by Natural Selection.

The entire emphasis of chapters Seven to Thirteen is that his theory, descent with modification through natural selection, explains a broad range of phenomena across many fields of inquiry and therefore manifests considerable consilience of inductions. The structure of Darwin's theory now looks like that shown opposite (illus. 9).

This is a robust theory. The axioms are jointly sufficient to allow the deduction that organisms will evolve and also to allow the deduction of a range of phenomena in numerous areas of biology. Those deductions provide powerful explanations of observed phenomena by demonstrating that they are to be expected given the mechanisms operating in nature, as captured in his axioms. The slightly more than 150 years that has elapsed since the publication of *The Origin* has refined and strengthened Darwin's theory. The essential elements outlined in the diagram remain the core of contemporary evolutionary theory. The refinement and strengthening of the theory from Darwin to the present are both evidential and conceptual. On the side of evidence, we have a much richer fossil record and better methods of dating. Stronger connections among fossils and geology and geomorphology have been forged. We also have a vastly better knowledge of embryological development, anatomy and the nature and functioning of cells. On the side of conceptual advances, a robust understanding of heredity at both the population and molecular level is in place. Darwin knew that traits were inherited but had no knowledge of the mechanisms of heredity. We now know the mechanisms by which new variations arise in populations. The concept of natural selection has been enriched by expanding its scope to capture more than a struggle for existence. There is a plethora of ways in which natural mechanisms bring about differential reproductive success among individual organisms. We now have an understanding of speciation (the conditions for new species to arise) and the mechanisms underlying it. A principal requirement for most speciation is, as Darwin claimed, isolation of sub-populations that once were part of a larger breeding group. In the chapters that follow we will explore these refinements and strengthening of Darwin's theory.

The essential features of Darwin's theory centre on the remarkably small variation of traits among organisms within a species, the advantage or disadvantage of specific traits in reproductive success, and the heritability of traits. Darwin held that selection acts on these small variations in traits and gradually, trait by trait, transforms the species. 'Gradualism' is the term commonly used to express that slow,

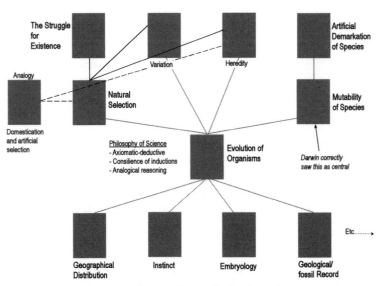

9 Darwin's axioms and the consilience of inductions; the top row gives the axioms.

incremental process. Darwin adopted a more general view of gradualism early in his career while reading Lyell's *Principles of Geology* during the *Beagle* voyage (27 December 1831–2 October 1836). Darwin, based on his observations during his *Beagle* travels, wrote convincing explanations of geological formations, which employed Lyell's gradualism. The biologist Thomas Henry Huxley (1825–1895), often called Darwin's bulldog and a strong advocate of evolution, did not accept his gradualist views; in a letter to Darwin dated 23 November 1859 (the day before *The Origin* was officially released to the public; it had been made available to booksellers on 22 November), he urged Darwin not to give gradualism such prominence.

This chapter has exposed the structure of Darwin's theory and its explanatory power, which is what makes it robust. It has also exposed the difficulties in it, which took about 60 years to resolve. For the most part the solutions confirmed Darwin's original theory and his claims.

THREE

Missing Pieces:
Heredity and Variation

The ink was barely dry on the first printing of *The Origin* when the debates began. There was a lot at stake for politicians, the clergy, the aristocracy and scientists. At the time the social structure in Britain was fragile. The spectre of political unrest had been constant since the American Revolution of 1776 and then the French Revolution of 1789; the unrest in France lasted until 1799. Other upheavals continued in Europe after that time, Germany (Prussia) being a prime example. The dissolution of liberalism and at least the semblance of democracy after 1848, under prime minister Otto von Bismarck's Realpolitik, provided a compelling example of what might happen in Britain.

The population was increasing dramatically in England: from 8.3 million in 1801 – the beginning of real census collecting – to about 17 million in 1851. By the 1861 census the population was more than 19 million. The city of London was stressed by population growth. It grew from 959,300 in 1801 to 2,363,000 in 1851. In the next 40 years it would increase to 5,572,012. This was partially caused by the yearly birth rate exceeding the death rate in London, but a significant portion of the increase was caused by an influx of workers, who were needed in the factories that were a consequence of the Industrial Revolution. The strains and tensions of this increase were causes for concern. Provision of housing, utilities (potable water and sewage, for example) and other services lagged well behind the growth in population. This kind of situation had been experienced before. The lesson learned was that it created a potential for discontent. The possibility of an uprising was always present.

To make things worse, the glue of British social cohesion – religion – was wobbling. Scientific evidence was challenging religious belief in many fields. This was seen most dramatically in geology. The geological evidence indicated that the biblically dated age of the earth (some 6,000 years old) was very wide of the mark. Something closer to millions of years was more consistent with the evidence. Fossils suggested that species had become extinct frequently over those millions of years. Even more alarmingly, the fossil record clearly indicated that humans had not existed in the earliest of those millions of years.

Added to the scientific challenges was a shift in confidence within the Church of England. For almost 2,000 years it had been confident that the Bible provided historical and scientific truths. Britain broke away from the Catholic Church in 1533 when Pope Clement VII refused to annul the marriage of Henry VIII and Catherine of Aragon. Nonetheless, the clergy in the newly formed Church of England and its parishioners had confidence that the Bible was trustworthy on matters of history and science as well as faith. Advanced biblical scholarship, which emerged first in Germany, changed that confidence. By 1859 many clergy in England had acknowledged that the Bible was neither history nor science. They even accepted that reconstructing a historical account of the life of Jesus was impossible.

In this fragile social setting, Darwin's theory was explosive. Its challenge to religious orthodoxy was all too clear. All life and especially humans, according to his theory, were the product of a totally natural process and not special creations of God. Darwin never extended his theory to humans in *The Origin*. 'Man' is mentioned only once. The implications for humans were obvious to everyone, however. One real and immediate threat was discerned: remove the religious basis for social organization in Britain and the principal instrument of social control evaporates. That consequence was seen as potentially catastrophic. In the context of the stresses of population expansion (crowding, housing, water and food) and the recent history of uprisings elsewhere, the loss of social cohesion was terrifying to those in power. Politicians, the clergy and aristocrats could lose everything they had, especially wealth and social power.

There was also reason for consternation in the scientific community. For one thing, scientists had built careers on what, if Darwin was correct, was a foundation of sand. It is therefore no surprise that they rejected his theory and mounted arguments against it. In addition resistance to change is widespread. Scientists are no different from the rest of the population. We have been inculcated into a way of thinking from birth. Hence many ideas seem correct and obvious, and radical digressions from those ideas can seem obviously wrong. Ironically, scientists were more entrenched in their views than the general population. Of course, they had a lot more at stake. Darwin's theory was a scientific one that directly challenged their scientific convictions. Careers had been built on a different view of things. The future would be uncertain in a world where that view was dismantled.

Debate within the scientific community focused mainly on four elements of Darwin's theory. First, there was a debate among those who accepted evolution. Many remained unconvinced by gradualism, Darwin's view that natural selection acting on small variations was the major mechanism of evolution. Thomas Huxley, one of Darwin's most ardent supporters, parted company with him on this point. Huxley held that evolution required major deviations within a group of organisms and that those jumps were the fuel for natural selection. This view is known as saltationism. In 1859, after reading *The Origin*, Huxley wrote to Darwin: 'You have loaded yourself with an unnecessary difficulty in adopting *Natura non facit saltum* [nature does not make jumps] so unreservedly.' Others held that natural selection acted on major differences in type. One obvious candidate for a major deviation from type (that is, from the norm for a species) is a 'sport', which frequently is caused by a genetic mutation. Darwin firmly rejected saltationism throughout the rest of his life. His careful study of artificial selection, where breeders selected small variations, provided what he considered clear evidence that natural selection acted on small variations. This debate was not resolved until the 1920s. During that decade, Mendel's and Darwin's theories were integrated. That integration allowed a proof that selection acting on small individual variations could lead to large changes over time. Indeed, that mechanism could lead to the creation of a new species.

The second debate focused on heredity. Heritability of traits is crucial to Darwin's theory but remained shrouded in mystery until the early twentieth century, when Mendel's experiments and theory became widely known; Hugo de Vries and Carl Correns replicated aspects of his experiments, which brought Mendel's work to prominence in 1900. By 1868 Darwin had adopted a view of heredity according to which every organ of the body sheds small particles (gemmules), which collect in the organs of reproduction and through which they are passed to the offspring. This view is known as pangenesis.

The French naturalist Jean-Baptiste Lamarck (1744–1829) held that environment impacts were heritable and that this was the principle mechanism of evolutionary change. The classic, although not completely accurate example, is the giraffe stretching its neck to reach higher on trees. This led to longer and longer and longer necks over many generations. Adapting Lamarck's idea of the 'inheritance of acquired characteristics', Darwin reasoned that since the environment could cause changes to the organs of the body, an environmentally changed organ would shed altered gemmules. This was the source of new variation in every generation. The details of how this pangenesis mechanism functioned were sketchy. Darwin never provided a comprehensive or well-formulated theory of pangenesis.

The third debate concerned the sources and maintenance of variation. These were not well understood and many, on statistical grounds, challenged Darwin's assumption that variation would continue to be extensive despite the action of selection. Some held that heredity and the generation of variation were opposing forces. Some believed that heredity itself generated variation. Darwin held that heredity resulted in traits blending from generation to generation (blending heredity). This he saw as consistent with his pangenesis theory. Without blending, Darwin held, the creation of abundant variation in every trait from generation to generation would render the persistence of any trait impossible. Any given trait would be swamped. Trait persistence is essential to evolution. Blending heredity is what tames the relentless creation of variability.

The fourth debate centred on the age of the earth and whether there had been enough time for gradual evolution. The Irish and

British mathematician and engineer William Thomson, Lord Kelvin (1824–1907), estimated, based on his calculation of the time the earth had taken to cool to its current temperature, that its age was between 20 and 400 million years; he thought 98 million years was most likely. This time is far too little for Darwin's gradual evolution. It was also inconsistent with the prevailing Lyellian principles of geology. In the view of Kelvin and his followers, that was a strike against Lyell's principles.

The last three of these four issues are the focus of this chapter. In the four decades after the publication of *The Origin*, inheritance and the sources of variation received the most attention. Despite all the attention, resolving these eluded everyone. The age of the earth turned out to be a minor skirmish by comparison, although an exceptionally important one. By 1900 all three remained unresolved. In that year Mendel's work began to be more widely known. One might think that would have settled the debate about the mechanism of heredity. Remarkably, however, his theory only fuelled more debates. But that story is for later chapters.

Throughout these debates and until his death, Darwin, for the most part, held to his original theory: natural selection acted on small variations, not 'sports'; those variations are heritable and new variations arise in each generation. The last two claims were based entirely on observation and not on known mechanisms.

For convenience, we can see the drama unfolding in three stages. The first stage begins with the publication of *The Origin* and ends in 1900. This is the stage covered in this chapter. The second stage covers the first decade of the twentieth century, spilling a little into the second decade. This period was marked by intense and unfriendly disagreements. Two camps had emerged by the end of the first period and the feuding between them began in earnest during the second. The feud was based as much on personalities as conceptual differences. The incipient phase of a third period began in 1910 but was not in full swing until about 1915 and ended in 1930. This is the period during which Darwin's theory was integrated with Mendel's theory and with biometrics, which is explored later in this chapter.

Some Prominent Characters, 1859–1930

Nine people figure prominently in the first two periods: Francis Galton, W.F.R. (mostly called Raphael) Weldon, Karl Pearson, William Bateson, Wilhelm Johannsen, Godfrey Harold (G. H.) Hardy, G. Udny Yule, Wilhelm Weinberg and August Weismann. Ronald Fisher, J.B.S. Haldane and Sewall Wright are the main characters in the third period. We shall get to know them a little better before exploring the roles they played.

Francis Galton was intellectually gifted, a fact that became obvious at an early age. Today Galton is known, if at all, for his advocacy of eugenics, a term he coined in 1883. Eugenics, however, comprised a small part of his intellectual interests and accomplishments. When he died in January 1911 at the age of 88, he had written 23 books and more than 400 articles, reports and reviews in fields as diverse as statistics, meteorology, audiology and forensics. The unifying thread of his work was the application of measurement (quantification) to these fields and the use of statistics. In 1909 he was knighted. He was a pioneer in applying statistics to biological phenomena and came remarkably close to Mendel's theory – based more on mathematics than experiments.

W.F.R. Weldon was born in London. His father, Walter Weldon, was a journalist and industrial chemist. In 1876 he entered University College, London, and planned to study medicine. While there he attended the lectures of the mathematician Olaus Henrici and the zoologist E. Ray Lankester. In 1877 he transferred to King's College, London. He left University College with a workable knowledge of mathematics. His knowledge was adequate for providing a mathematical account of zoological facts but far from a deep understanding. In 1878 he moved again, this time to St John's College, Cambridge. He arrived at Cambridge in April and continued his zoological studies, working with Professor Francis Maitland Balfour, who was a renowned embryologist – indeed, one of the fathers of the field.

Medicine was no longer Weldon's goal. He was now fully committed to zoology. In 1881 he graduated with first-class honours in natural science. In 1882, after spending a year at the Stazione Zoologica di Napoli, he secured a lectureship in invertebrate morphology. In 1889 he became Chair of Zoology at University College, London. His work focused on crabs and shrimps. His skill at sectioning and staining cells was excellent. So was his ability to draw what he saw through the microscope. Shortly after its publication in 1889, Weldon read Francis Galton's *Natural Inheritance*. Weldon's competent but not deep knowledge of mathematics allowed him to see immediately the relevance of Galton's frequency distribution work (discussed later) to his own studies. He commenced experiments on shrimps to apply Galton's distributions. In 1890 he was elected to the Royal Society.

Karl Pearson (born in London on 27 March 1857) received his university education at King's College, Cambridge, where he studied mathematics. Between 1878 and 1880 he also studied medieval and sixteenth-century German literature in Berlin and Heidelberg. The University of Cambridge offered him a position in its German department, which he declined. He then studied law. He was a person of great intellect with many talents and interests.

In 1881 he served as deputy professor of mathematics at King's College, London, and then, in 1883, at University College. In 1884 he was offered the Goldshmid Chair of Applied Mathematics and Mechanics at University College, which he accepted. In 1891 he accepted the professorship of geometry at Gresham College. Although we get a little ahead of our story at this point, it was at Gresham that he met Weldon. Thus began a remarkably fruitful collaboration. It was through Weldon that Pearson met Francis Galton and a triumvirate collaborative team was formed. He was not the best mathematician of his era but he was very good. He was near the top of his class at Cambridge, and was third wrangler: that is, he was third among those awarded a first-class degree in the three mathematics exams. He made a number of notable contributions to statistics and was only 28 when he became Chair of Applied Mathematics and Mechanics at University College, London. He was intolerant of incompetence and came to view William Bateson as utterly incompetent, especially in mathematics. This surprised no one, since mathematics was Bateson's Achilles heel.

Pearson also was active as a socialist. He lectured on women's rights and, controversially, on Marxism. In 1920 he was offered an Order of the British Empire, which he refused, based on his socialist principles, and in 1935, for the same reason, he refused a knighthood.

William Bateson (born in Whitby, Yorkshire, on 8 August 1861) was not a stellar student and mathematics was his major weakness. His principal success was in natural sciences, which he studied at St John's College, Cambridge. At that time, Weldon was also at St John's and he and Bateson became good friends. Bateson did careful embryological work in the tradition of the animal morphologist and embryologist Francis Balfour under whom he, along with Weldon, studied at Cambridge. Both Bateson and Weldon soon moved beyond Balfour's views and methods. In 1883 Bateson came under the influence of W. K. Brooks, who had a keen interest in variation and evolution and in 1883 published *The Law of Heredity: A Study of the Cause of Variation and the Origin of Living Organisms*. He claimed that his theory of heredity was superior to Darwin's pangenesis theory. His view embraced the saltationist account of natural selection. The influence of Brooks coupled with evidence Bateson gathered on a research trip to Russia led him to conclude that there was no evidence to support Darwin's contention that natural selection acts on small individual differences. This confirmed for him the correctness of Brook's saltationist view.

It is reasonable to credit Bateson with being the father of the science of genetics and he gave it its name. His enduring contributions were few. Mostly they flow from his experiments, which provided evidence basic to the modern understanding of heredity. He was a committed evolutionist but not a Darwinian. Indeed, he came to disagree dramatically with Darwin's theory. As noted, by 1894 he became convinced that evolution had not, and indeed could not, occur through selection of small individual variations. Evolution needed to be discontinuous. He also parted company with Darwin on variation.

To bolster his views, he began to focus on experiments involving the breeding of plants and animals. Having founded the field, he became the University of Cambridge's first professor of genetics in 1908. In 1910 he moved to the position of director of the John Innes Horticultural Institution at Merton, south London.

Like a few others, in 1900 Bateson became aware of Mendel's paper on plant breeding and the theory of heredity based on those experiments. He translated Mendel's paper into English and was a vigorous advocate of his theory. As our story unfolds, it will become clear that he was on the wrong side. Darwin's views on natural selection triumphed. Nonetheless, Bateson made important contributions to genetics. He discovered that certain traits were inherited together; this seemed contrary to what Mendel's theory appeared to predict. Bateson turned out to be correct and the phenomenon was termed linkage. It became clear later that it was entirely consistent with Mendel's theory. Linkage occurs when genes are close together on the same chromosome. As the term implies, these genes are linked. He also showed experimentally that some traits are controlled by more than one gene.

Wilhelm Ludvig Johannsen was born on 3 February 1857 in Copenhagen, Denmark. During his early years he worked as an apprentice pharmacist, and qualified after passing the licensing exam. He never attended university; his father lacked the resources to fund him. Hence his academic career began in 1881 when he secured a position at the Carlsberg Laboratory, which had been founded in 1875 by J. C. Jacobsen, who earlier had founded the Carlsberg brewery. The laboratory had departments of chemistry and physiology; Johannsen was appointed an assistant to Johan Kjeldahl in the department of chemistry. Kjeldahl's work focused on plant physiology. Johannsen's main discovery during this period, drawing extensively on his pharmaceutical background, was that dormant seeds could be made to germinate using ether and chloroform.

Having established a significant reputation in plant physiology, he moved in 1892 to the Royal Veterinary and Agricultural University to take up a lectureship. He was promoted to professor at that institution in 1903. He coined the terms 'gene', 'genotype' and 'phenotype', and developed an experimental technique for tracing and creating through many generations what he termed 'pure lines'. Johannsen's first book became a widely used textbook of plant genetics. As a result he was sought after as a lecturer in Europe and North America. His interpretations of his experimental results suggested that Darwin and his followers (Weldon, Pearson in particular) were wrong about variation and selection. In 1905 Johannsen left the university to become professor of plant physiology at Copenhagen University; he was made rector of that university in 1917.

In 1924 Johannsen became a member of a state commission on degeneration and eugenics, although he never fully embraced the eugenics movement. He spent his later years writing on the history of science and epistemological aspects of the new genetics (epistemology is the study of the nature of knowledge and of knowledge acquisition). His research and philosophical commitments led him to reject genetic determinism, the view that nature is the dominant cause of traits; the environment plays a minor role. He argued for a significant role for the environment and that this duality of genetics and environment rendered indeterminate claims that mental illnesses, alcoholism and a range of diseases were heritable traits. He also held, based on his experiments, that environmental impacts on the phenotype (adult organism) cannot influence the genotype (genes), making him a staunch anti-Lamarckian.

Although the interpretations of the results of his research were later rejected, at the time it was considered brilliant and groundbreaking work. As became a pattern, Darwin's views as expressed in *The Origin* turned out to be correct and his critics turned out to be wrong. Johannsen is a classic example.

Godfrey Harold Hardy was mostly known as Hardy; some close friends called him Harold. He detested the name Godfrey and all his mathematical publications were published under the name G. H. Hardy. He was born on 7 February 1877 in Surrey, England, and from an early age was regarded as exceptionally brilliant, a fact he embraced. From an early age he was shy, disliking public attention, photographs and mirrors. He was also somewhat frail.

Although Hardy had only a minor interest in mathematics as a boy, he became an exceptional mathematician. He began his university studies at Cambridge in 1886. He was fourth wrangler in the mathematical tripos (course) in 1898, a placing that irritated him for years. Given his subsequent career and contributions to mathematics, his irritation seems entirely justified. The famous mathematician David Hilbert, in a letter, referred to him as the best mathematician in England. His academic career began at Trinity College, Cambridge. He was elected to the Royal Society in 1910 and received its Copley Medal in 1947. In 1919 he accepted the Savilian Chair in Geometry at Oxford University.

Hardy was a fan of cricket, walking almost daily to the university cricket pitch in the afternoon and reading the cricket scores in the newspaper early every morning. His mathematical work was done from around 9 am to 1 pm almost every day. He was considered eccentric by his friends and colleagues. His importance to our story is that in 1908 he and Wilhelm Weinberg, simultaneously but independently, published a crucial equilibrium principle for Mendelian genetics.

Wilhelm Weinberg was born in Stuttgart, Germany, on 25 December 1862. He obtained his MD in 1886 and set up practice as a general physician and obstetrician. He is reported to have attended more than 3,500 births during his lengthy career.

The fact that both independently discovered the equilibrium principle, which is described in more detail later, resulted in it being named the Hardy-Weinberg equilibrium. Notwithstanding this link, in many respects the two were very different. Unlike Hardy, who was a mathematician and whose foray in biology was fleeting, Weinberg was a German physician (obstetrician and gynaecologist) who went on to make many contributions to biology. Also unlike Hardy, who disliked applied mathematics, Weinberg saw value in it for medicine. This can be seen in his publications from the beginning of his career. For example, in his 1901 paper 'Beiträge zur Physiologie und Pathologie der Mehrlingsgeburten beim Menschen' – an 85-page volume – he developed and used the difference method. His interest was in the proportion of monozygotic twins (those resulting from a single fertilized egg that at some point divides into two separate individuals – usually referred to as identical twins) and dizygotic twins (those resulting from two separate fertilized eggs) at birth. He used statistical data on the sex of twin births, where no monozygotic or dizygotic differentiation had been made in the data. He examined that data using his difference method. This allowed him to determine the proportion of monozygotic to dizygotic twins.

Weinberg mostly worked alone and his work had only a minor impact in Germany at the time and none in Britain. He read a paper on his equilibrium principle on 13 January 1908 to the Society for the Natural History of the Fatherland in Württemberg. It was another six months before Hardy's paper appeared in the *American Weekly*

Science, in July 1908. Weinberg's lecture was published later in 1908 in the society's yearbook. For many years the principle was known as the Hardy Equilibrium. It would be 35 years before Weinberg's work became known in the English-speaking world and his name was attached to it.

G. Udny Yule was born on 18 February 1871 in Morham, Scotland. He began his university studies in 1887 at University College, London. In 1890 he was awarded a degree in engineering, a profession he quickly came to dislike. Consequently he turned his attention to physics. His engineering degree gave him a good basis for this new area of study, but a year in Bonn studying physics told him that this was not the field for him either. It was serendipitous that he ended up as a statistician. Karl Pearson, who knew Yule from student days, was creating his fledgling statistical group at University College and saw Yule as a good fit. Yule was offered a position as demonstrator. His *Introduction to the Theory of Statistics* of 1911, a field in its infancy, rocketed him to fame in the statistical and scientific world. In 1912 he was offered the newly created lectureship in statistics at Cambridge (roughly the equivalent academic rank as the North American position of assistant professor). In due course, he was promoted to reader (similar to the North American position of associate professor). During the First World War, he worked for the Ministry of Food. After the war he returned to St John's College, Cambridge. After a stellar career, he retired in 1930; by then the field of statistics was well established. Yule made significant contributions to:

1) The theory of correlation (i.e. the degree to which two or more things – traits, events and the like – are related. Two things that are completely independent have a correlation of 0 and two things that always vary together and to the same degree have a correlation of 1).

2) The theory of partial correlation.

3) The theory of linear regression for multiple variables (linear regression involves finding the straight line that best represents the pattern of an array of data; see illus. 10), and

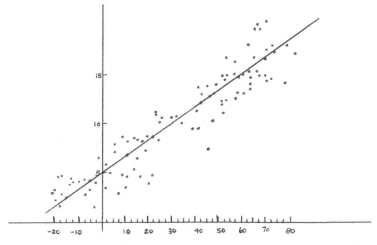

10 Linear regression: finding the line such that the data points (the dots) above and below are equal. The line gives 'the best straight line – linear – fit to the data'.

4) Features of the binomial distribution (discussed in a little more detail later in this chapter).

Yule was a brilliant statistician and was showered with honours for his outstanding contributions. He was elected to the Royal Society in 1922, was awarded the gold medal of the Royal Statistical Society in 1911 and received numerous other domestic and international honours. As the British statistician M. G. Kendall wrote in the *International Encyclopedia of Social Sciences* in January 1968,

Yule's outstanding contribution to statistics results not so much from any one quality as from his combination of qualities. He was not a great mathematician, but his mathematics was always equal to the task. He was not trained in economics or sociology, but his wide knowledge of human relationships enabled him to write with insight on both subjects. He had the precision, the persistence, and the patience of a true scientist but never lost sight of the humanities. He was a kindly, genial, highly literate, approachable man who refused to embroil himself in the controversies that mar so much of statistical literature. Above all, he had the flair for handling numerical data that characterizes the truly great statistician.

August (Friedrich Leopold) Weismann was born on 17 January 1884 in Frankfurt, Germany. He is widely regarded as one of the most significant evolutionary biologists of his time and his work has had lasting importance. His most important contribution was the germ plasm theory. Biologists distinguish between somatic cells (the cells that make up organs of the body – the skin being the largest organ) and what Weismann called germ cells but today are called gametes (egg (ovum) and sperm). Weismann's germ plasm theory stipulates that it is only through gametes that hereditary transmission occurs. Weismann studied medicine instead of his subject of interest, the natural sciences, largely because of his parents' perception of the poor job prospects in natural sciences. He graduated in 1856. Then his career took off. He won two prizes for essays: one on hippuric acid in herbivores, the other on the salt content of the Baltic Sea. In 1861 he read *The Origin*, reportedly from beginning to end in one sitting, and became a committed Darwinian. For a brief period he practiced medicine – including being a field doctor in Italy, where he met his wife Maria Dorothea Gruber, and for about two years as the personal physician to Archduke Stephan of Austria. He then turned his attention to natural science and after 1863 he poured his intellectual energies into that field.

Weismann held the first chair in Zoology in the University of Freiburg. He was fully committed to Darwin's mechanism of natural selection, taught it to generations of students and vigorously defended it. He rejected, however, Darwin's theory of pangenesis. He also rejected Darwin's acceptance of a form of Lamarckism. In short he thought Darwin was entirely wrong about heredity and set out to develop his own theory of heredity. Though Mendel laid the foundation for understanding the dynamics of it, Weismann's theory laid the foundation for its actual biological processes. He was elected to the Bavarian Academy of Sciences, the Royal Society (London) and the Linnaean Society. Weismann travelled widely and frequently and, understandably, often to Italy. He visited England several times but surprisingly never met Darwin, with whom he had corresponded; Darwin was somewhat reclusive, especially in his later years.

Ronald Aylmer Fisher was born on 17 February 1890 in London. He studied mathematics and astronomy at Cambridge from 1909 to 1912, when he graduated. As a student, he was active in the eugenics movement, founding the Cambridge University Eugenics Society. He later became Galton Professor of Eugenics at University College, London. It was during his undergraduate years at Cambridge that he became interested in evolutionary theory.

Fisher was an outstanding mathematician. His most notable mathematical contributions were in statistics. After he graduated, he began work on observational error theory. From 1914 to 1919, through the war years, he taught secondary-school mathematics and physics. His hours outside the classroom were devoted to statistics and genetics. The year 1919 marks the beginning of his experimental work in agriculture. That year he joined the staff of the Rothamsted Experimental Station. Although his main role was as a statistician, he carried out experiments related to plant genetics.

In addition to contributions in pure statistics, he made contributions in applied statistics such as analysis of variance (the variance of a data point from the mean). Another of his contributions has become entrenched in experimental design in clinical medicine: randomized controlled trials. In his *Design of Experiments* (1935) he set out what he deemed to be a necessary method in order to uncover cause-and-effect relationships. The three critical elements are blocking, randomization and replication. He recommended the use of this method for a wide array of research domains. His application of this method was in agricultural research. In this context blocking was the technique of taking an experimental plot and dividing it into equal-sized small areas of the plot (blocks). One of two adjacent blocks would randomly be designated the intervention block. The other is the control. For example, the intervention block might have fertilizer added. The other would have no treatment. Fisher argued that if there was a difference in, say, the vigour of growth of the plants in the intervention block compared to those in the control (non-treated block), then fertilizer can be declared the cause of the vigorous growth.

His absolute conviction that this method was the only way to discover causes put him on the wrong side of history in the debate over smoking and lung cancer. Richard Doll and A. Bradford Hill published a landmark paper in the *British Medical Journal* on 30 September 1950. They examined data regarding carcinoma (cancer) of the lung, stomach, colon or rectum. The data was collected from patients at twenty hospitals between April 1948 and October 1949. They reported a clear 'causal' link between smoking and lung cancer. Fisher dismissed their research and findings because it did not employ a randomized controlled trial. He stated his position crisply:

> Seven or eight years ago, those of us interested in such things in England heard of a rather remarkable piece of research carried out by Dr Bradford Hill and his colleagues of the London School of Hygiene . . . *correlation is not causality.* The fact that if two factors, *A* and *B*, are associated – clearly positively, with statistical significance, as I say – it may be that *A* is an important cause of *B*, it may be that *B* is an important cause of *A*, it may be that something else, let us say *X*, is an important cause of both . . . [In the case of smoking and lung cancer, for] my part, I think it more likely that a common cause supplies the explanation. Again, we do not know. I do not put forth any explanation as proved, but as requiring investigation. The obvious common cause to think of is the genotype.

Fisher had suggested that a person's genetic makeup might predispose them to smoke and also predispose those same persons to develop lung cancer. For him, only a randomized controlled trial would uncover the real cause and effect, though he did recognize that such a trial would be unethical.

Fisher was crusty and egocentric, but his contributions to statistics, agricultural research and evolutionary theory are impressive. On the last topic, he was on the right side of history. Indeed, his work and conclusions were a pivotal contribution to the theory.

John Burdon Sanderson Haldane was born on 1 December 1892 in Oxford. Of our cast of early twentieth-century characters, he is unusual. He revelled in combative debates but was also convivial and nurturing. He saw no reason why one could not hold two conflicting views simultaneously, and he often did. He was a dedicated Marxist, rejecting with vigour capitalism and imperialism. He fought on the Front in the First World War, in the Scottish Black Watch, and reported that to his surprise he enjoyed killing Germans. His Marxism led him to become engaged in the Spanish Civil War on the side of the Republicans. He frequently experimented on himself, sometimes taking dangerous risks. He wrote a popular science column in the *Daily Worker* as well as a number of popular science books. At the age of 65 he left University College, London, and moved to India, where he became a citizen, adopted Indian garb, promoted the study of science and headed the government's Genetics and Biometry Laboratory. When near the end of his battle with cancer, in 1964, he wrote in the *New Statesman*:

Cancer's a Funny Thing:
I wish I had the voice of Homer
To sing of rectal carcinoma,
Which kills a lot more chaps, in fact.
Than were bumped off when Troy was sacked
Yet, thanks to modern surgeon's skills,
It can be killed before it kills
Upon a scientific basis
In nineteen out of twenty cases . . .

His wit and attachment to the humanities remained to the very end. Above all this complexity of personality, however, he had a remarkable intellect.

His father, John Scott Haldane, was a professor of physiology at Oxford University. J.B.S. Haldane assisted him with experiments from an early age. He bred guinea pigs while still quite young and, as a result, learned Mendelian genetics. He went up to Oxford in 1911 and graduated in mathematics, and in classics and philosophy. He held positions at Oxford, Cambridge and University College, London. At the last he occupied the newly founded chair in genetics. Haldane was both a theoretician, applying his mathematical prowess to scientific problems, and an experimenter. One of his enduring contributions as an experimenter was demonstrating genetic linkage – a tendency for genes that are close to each other to be inherited together. This is contrary to Mendel's principle of independent assortment. Hence Haldane's experimental results required an explanation, which itself required an understanding of the behaviour of chromosomes. One of his lasting contributions to scientific theorizing was his role in unifying Mendel's theory with Darwin's theory of evolution.

Sewall (Green) Wright was born on 21 December 1889 in Melrose, Massachusetts. He died on 3 March 1988, just short of his hundredth birthday. He learned basic arithmetic from his mother at an early age and, under her tutelage, he was reading and writing proficiently by the age of six. He was a precocious child who had an interest in how things worked. When he was seven he wrote a pamphlet titled 'The Wonders of Nature'. He had an excellent quantitative ability and looked for quantitative solutions to natural phenomena. Wright's father taught him poetry and his mother natural history and arithmetic. He was not stimulated by his elementary school education. Secondary school, however, offered a much more stimulating environment. Even though there were no courses in chemistry or zoology, he did study physics, in which his quantitative abilities flourished. He also studied algebra and geometry. He read voraciously but also engaged in extracurricular activities. He graduated from secondary school in 1906 with a very high average.

In 1906 he entered Lombard College in Galesburg, Illinois. Although Wright's studies had a more mathematics focus, he continued to be fascinated by natural history. During his first year at Lombard College he spent much of his spare time birding. The year before his senior year was spent working for the railroad. In his senior year he began to work with Francis Bruté Key (née Wilhelmine Entemann Key), who received her PhD in 1901 from the University of Chicago. She had studied with some of the leading biologists of her day. Wright's interest in biology, and his biological knowledge, exploded under her teaching. He read most of the influential works in evolution under her guidance. In hindsight R. C. Punnett's *Encyclopaedia Britannica* entry on Mendelism was pivotal. It included the Punnett square, which displayed the Mendelian ratios clearly. The Punnett square was used long before Punnett but his use was especially important for the clarity it allowed him to bring to mathematical issues. The Punnett square is a 2 × 2 display. Hence, there are four squares; the table on p. 84 is an example. Punnett's entry launched Sewall Wright's career in biology, especially genetics.

Wright graduated from Lombard College in 1911 and went, for the summer, to Cold Spring Harbor – a famous u.s. research centre. It was, and indeed still is, constantly visited by internationally renowned biologists. Understandably this was another transformative experience. He entered the University of Chicago in the autumn of 1911. He spent the spring of 1912 at the Agricultural College of the University of Illinois. It was there that he met William Ernest Castle, a professor of zoology at Harvard University. Castle was very impressed with Wright and agreed to supervise his doctorate. Wright completed a fast-track MA and spent another summer at Cold Spring Harbor. Then he was off to Harvard. His career was secured and his commitment to biological research, especially genetics, was sealed.

Although we will encounter a number of other contributors to the development of evolutionary theory as well as an understanding of the actual evolution of organisms, these nine were pivotal and had a high profile. Having met the cast, let us move on to the debates sketched earlier:

1) Was evolution gradual, as Darwin claimed, or discontinuous (making jumps)?

2) What is the mechanism of heredity?

3) What are the sources of variation; how is it maintained?

4) Was the earth old, as Lyell's geology maintained, or relatively young – too young for gradual evolution?

This chapter traces the debates during the first period (1859–1900).

In 1867 a review of *The Origin* was published in the *North British Review*. All reviews in that journal were anonymous, but at some point between June 1867 and early 1869 Darwin had somehow become aware that Fleeming Jenkin, a professor of engineering, was its author.

Jenkin develops three arguments against evolution. First, he argues that transmutation of species is impossible and Darwin's evidence for his view that species are a human construct is insufficient. The argument on this point is fragmented in the review, with some elements provided at the beginning and some later. Second, he argues that variations from type, in any generation, will be rare and that they will be eliminated in a few generations. Third, the earth is too young for Darwinian evolution.

The impossibility of transmutation argument is not unique to Jenkin, though he gives it one of its best formulations. Moreover, Darwin's arguments in favour of transmutation are more complicated than Jenkin seemed to appreciate. Darwin paid little attention to this argument. The argument regarding the rarity of variations from type and the high improbability of their being preserved caught Darwin's attention largely because it laid bare in a mathematical way what he already believed: individual differences, not variation from type, were the grist for natural selection. He had already rejected Huxley's view that 'sports' – significant variations from type – were the material on which selection acted. Jenkin demonstrated that such 'singular' individuals would not survive the reproductive processes and soon would be eliminated. Jenkin's arguments on the age of the earth also attracted Darwin's attention.

On 22 January 1869 he wrote to Alfred Russel Wallace:

I have been interrupted in my regular work in preparing a new edition of the 'Origin,' which has cost me much labour, and which I hope I have considerably improved in two or three important points. I always thought individual differences more important than single variations, but now I have come to the conclusion that they are of paramount importance, and in this I believe you agree with me. Fleeming Jenkin's arguments have convinced me.

In a subsequent letter to Wallace on 2 February Darwin is explicit about the connection of Jenkin and the *North British Review* article:

F. Jenkin argued in the 'North British Review' against single variations ever being perpetuated, and has convinced me . . . I always thought individual differences more important; but I was blind and thought that single variations might be preserved much oftener than I now see is possible or probable.

Darwin added a section in the fifth edition of *The Origin* (1869), which incorporates Jenkin's criticism. He does not mention Jenkin by name, although he is clearly addressing the *North British Review* article.

It should be observed that, in the above illustration, I speak of the slimmest individual wolves, and not of any single strongly marked variation having been preserved. In former editions of this work I sometimes spoke as if this latter alternative had frequently occurred. I saw the great importance of individual differences, and this led me fully to discuss the results of unconscious selection by man, which depends on the preservation of the better adapted or more valuable individuals, and on the destruction of the worst. I saw, also, that the preservation in a state of nature of any occasional deviation of structure, such as a monstrosity, would be a rare event; and that, if preserved, it would generally be lost by subsequent intercrossing with ordinary individuals. Nevertheless, until reading an able and valuable article in the 'North British Review' (1867), I did not appreciate how rarely single variations, whether slight or strongly marked, could be perpetuated.

These comments focus on the ineffectiveness of natural selection in preserving a variation from type because subsequent interbreeding in the population would eliminate it, what some historians have called Jenkin's swamping argument: a variation from type would be swamped in the breeding process.

Darwin had always focused on individual differences, which, although small, were found in sizable portions of individuals in a population. Some individuals are fast, some slow, many average in locomotion. If, in an environment, being fast is advantageous, then over time speed of movement will increase. Darwin was not mathematic-ally literate and indeed distrusted mathematical calculations when they were inconsistent with his observations. If he had resorted to math-ematics, as Jenkin did, he could have characterized his view in terms of a normal distribution of traits, something Francis Galton, Darwin's half-cousin (Erasmus Darwin was grandfather to both), was explor-ing during this period, as we shall soon see.

Jenkin's third argument was more problematic for Darwin's the-ory – as well as for Lyell's geological theory. William Thomson's (later Lord Kelvin) first estimate of the age of the earth, in 1864, came after the publication of the third edition of *The Origin* (1861). The extreme upper limit of his calculation was about 400 million years. His calculation was based on the cooling of the earth. Thomson was a mathematical physicist. He used thermodynamic principles to determine the rate of cooling of the earth to arrive at his range of its age. His method, his data and his conclusions were criticized by most geologists, as well as John Tyndale and Thomas Huxley. Tyndale was an accomplished physicist whose criticism was not to be ignored. Thomson chose to ignore it. Huxley was a brilliant anatomist and evo-lutionist. Despite these criticisms, Kelvin's dating of the age of the earth is important because it is explicitly cited in a critical review that Darwin took very seriously.

If the earth was less than 400 million years old, there had been insufficient time for the evolution of the array of living things pres-ent in 1859. Darwin does not dwell on this point, perhaps because there was little to be said except that Lyell's geology and his theory of evolution provide better naturalistic explanations of the observed world than any alternatives. Two possibilities existed: either evolution at some earlier time occurred more rapidly than at present or Thomson's calculations must be missing some key elements. Darwin flirted with the latter, suggesting at one point that electromagnetic forces might affect Thomson's calculations based on heat. He abandoned that view when the Scottish scientist James Croll assured him that such forces could at best play an infinitesimal role, if any at all. Croll also assured Darwin that Thomson's calculation of the age of the earth was sound, writing:

If there is one thing more than any other in physics, regarding which we have absolute certainty, it is that the solar system is losing its store of energy. We not only know this fact, but we have a means of determining the actual rate at which it is losing its power.

Croll was self-taught in physics and not really up to the task of confirming Thomson's methods or conclusions. Moreover, this was not the area in which he concentrated his efforts. His contributions were in the field of celestial causes of climate change (ice ages, thaws and the like). He was also interested in geological effects on climate change and the effect of climate changes on geology. In this area he made some important contributions. Importantly, he corresponded with Lyell on geological matters and he also corresponded with Darwin. Although Darwin trusted him on knowledge and methodology in physics, he was never entirely convinced.

Faced with this 'absolute certainty', Darwin contemplated a faster pace for evolution. One mechanism to which he turned was sexual selection: the role of mate selection and the evolution of physical and behavioural traits in males, traits that were designed by selection to attract females. This form of selection, he thought, might speed up the selective process. In *The Origin*, he gave it a scant two pages, commenting:

This form of selection depends, not on a struggle for existence in relation to other organic beings or to external conditions but on a struggle between the individuals of one sex, generally the males, for possession of the other sex. The result is not death to the unsuccessful competitor, but few or no offspring. Sexual selection is, therefore, less rigorous than natural selection.

In 1871, in *The Descent of Man and Selection in Relation to Sex*, he devoted more than 500 pages to sexual selection (dealing with both human and non-human organisms); he began his treatment with an almost 70-page exposition of the principles of sexual selection.

Darwin also allowed for a faster pace in the early period of evolution by relaxing slightly his commitment to uniformitarianism. This is the view, championed by Lyell, that the forces of nature that were at work in the past were the same in kind and magnitude as those at work in the present. There were no special forces at work in the past and no catastrophes that cannot be explained by appeal to forces known to be at work today. This was the view Lyell advocated in his *Principles of Geology* and that Darwin adopted. In the fifth edition of *The Origin*, published on 7 August 1869, he added material on Thomson's calculations (misspelling his name) and acknowledged Croll's support of them. He clearly remained sceptical, writing (emphasis added):

> Sir W. Thompson concludes that the consolidation of the crust can hardly have occurred less than 20 or more than 400 million years ago, but probably not less than 98 or more than 200 million years. *These very wide limits show how doubtful the data are; and other elements may have to be introduced into the problem.*

Darwin clung to the hope that other elements would change the calculations and lengthen significantly the age of the earth. He turned out to be correct. Thomson's and Croll's confidence was entirely unfounded. (This would not be the only time that Thomson's cockiness was misplaced. For example, in 1895, he confidently claimed: 'Heavier-than-air flying machines are impossible.' In 1902, just seven years later, Orville and Wilbur Wright successfully glided in the air. In 1903 Orville Wright flew 20 ft off the ground in a powered, heavier-than-air machine. In 1900 Thomson, then Lord Kelvin, stated in an address to the British Association for the Advancement of Science: 'There is nothing new to be discovered in physics now. All that remains is more and more precise measurement.' Five years later, Albert Einstein published his Special Theory of Relativity and transformed physics.)

Notwithstanding his scepticism, Darwin thought it prudent to entertain a weakening of his commitment to uniformitarianism in order to increase the rate of evolution during its early period. In the sixth edition of 1878, he wrote:

It is, however, probable, as Sir William Thompson insists, that the world at a very early period was subjected to more rapid and violent changes in its physical conditions than those now occurring; and such changes would have tended to induce changes at a corresponding rate in the organism that then existed.

Although with this concession Darwin conceded that forces with a magnitude not found today were at work for some brief and early period, he did not concede that these were non-natural forces. The main thrust of Jenkin's arguments was that appeal to special creation was required to explain the array of living things. Darwin rejected such supernatural explanations. In *The Origin*, he wrote: 'It is so easy to hide our ignorance under such expressions as the "plan of creation," "unity of design" etc., and to think we give an explanation when we only restate a fact.'

Darwin considered Jenkin's review important because it sharpened his rejection of deviations from type as the focus of natural selection and because it made him wrestle with the time available for evolution. Although Jenkin's arguments against the mutability of species were unoriginal, his exposition of them was better than most. Nonetheless, Darwin correctly believed that he had dealt with them. Remarkably, many biblical literalists in the u.s. still put them on display as part of their refutation of evolution, although few know about their origin or Jenkin's formulation of them. Biology, and science generally, has moved well beyond this and the evidence for mutability, as we shall see, is now overwhelming.

Heredity was a different kind of challenge. There was no controversy about its occurrence. The challenge was that no clearly acceptable mechanism had been given. Without such a mechanism it was not entirely obvious that natural selection acting on small individual variation would lead to selection. The observed resemblance of offspring to parents did not guarantee that small differences were inherited consistently enough for evolution to be the result of Darwin's version of natural selection. Francis Galton greatly admired Darwin and accepted the fact of evolution but rejected Darwin's commitment to gradualism. He was also sceptical about pangenesis, for which there was no available empirical evidence. Galton was interested in exploring just what experimental evidence might show. With Darwin's encouragement, he undertook several experiments to test the theory. Specifically, he set out to test whether gemmules circulated in the blood. Darwin supported and encouraged Galton. This was quite reasonably taken by Galton to mean that Darwin believed that blood was the medium of transportation for the gemmules.

Galton transfused blood from mongrel into pure-bred rabbits. No effects on the offspring were found. He thought that it might be his removal of fibrin (a clotting factor in blood) before transfusion that was at fault. Hence he devised a method for making a direct exchange of blood through the carotid arteries of the rabbits. The offspring still failed to manifest any effect. These results were inconsistent with blood as the medium for the transport of gemmules. In response Darwin simply dropped the idea, indicating that he had never been committed to blood as the medium. He claimed that he had only ever written that gemmules were 'circulating freely' in the body and offered no other medium of transportation.

As a result of his experiments, Galton was confident that pangenesis was either wrong or needed to be significantly modified, although he knew his experiments were not conclusive proof of this. Hence he began his quest for an alternative theory. That quest was conceptual and mathematical, not experimental.

The postulate of Darwin's pangenesis that Galton found entirely untenable was the inheritance of acquired characteristics. Darwin had tried to explain why amputations did not have any effect on the next generation in terms of the environmental influences, which are able to change the gemmules. There was also the matter of reversions, in which a trait of an ancestor reappears several generations later. Darwin attributed this to dormant gemmules:

> These granules may be called gemmules. They are collected from all parts of the system to constitute the sexual elements, and their development in the next generation forms a new being; but they are like-wise capable of transmission in a dormant state to future generations and may then be developed.

Galton was not convinced.

Galton's first exposition of his alternative theory was published in the *Proceedings of the Royal Society* in 1872 under the title 'On Blood-relationship'. Unfortunately, it was underdeveloped and verged on incoherence. In 1875 Galton produced a popular, more accessible and more developed account of his theory. It was published in the *Contemporary Review* under the title: 'A Theory of Heredity'. He introduced a new term, 'stirp'. A stirp is 'the sum total of the germs, gemmules, or whatever they be called, which are to be found, according to every theory of organic units, in the newly fertilised ovum'. Although this sentence expresses some indifference about terminology, he preferred 'germs' to 'gemmules'. This was mostly to distinguish his understanding of the hereditary units from Darwin's. The ovum, he tells us, contains the stirp as well as required nutrients. In contemporary terms his stirp is similar to a nucleus in a cell.

He then set out four postulates, which he claimed were uncontroversial. In capsule form, Galton's theory was that there is a separate germ for every unit of the body. One can assume these 'units of the body' are cells. There are, however, times when the phrase 'unit of the body' refers to a collection of cells, such as an organ. There are more germs than are required for determining the structure of the adult organism. That is, not all the germs in a stirp are expressed in any specific organism. Those germs that do not participate in the development of a specific organism retain their potential for doing so in the development of another organism – a sibling or offspring, for example. This is the explanation of reversion: traits showing up again several generations later. Construction of the adult organism is a dynamical one, with germs attracting and repulsing each other.

This is the foundation of Galton's theory but he knew that more was required – principally some method of reduction. If germs from a male and female combine to create a stirp, there will be double the number in the newly formed stirp than in either parent. In just a few generations the number of germs in a stirp would have ballooned. In true Darwinian fashion, he thought that the reduction resulted from a struggle for existence. Half the germs would lose in this struggle. Hence natural selection occurs at the stirp level and not just at the organism level. The theory is ingenious, but not very sophisticated or well worked out.

So far Darwin would find little to which to object in the theory. The point of divergence occurs in the second half of the paper. It centres on how the germs in a stirp behave and how stirps arise. Here, he introduces yet another term: 'septs' (divisions). Stirps are composed of septs and septs are composed of germs. This is becoming a bit complex, but essentially he was rejecting pangenesis. So that his readers would not miss this fact, he stated it explicitly: 'This is a very different supposition to that of the free circulation of gemmules in Pangenesis.'

Galton had, rather inelegantly, veered in the direction of the still-unknown theory that Mendel had quite elegantly set out ten years earlier. Contrary to Darwin, but like Mendel, Galton separated the sexual cells (egg and sperm) from cells making up the rest of the organism. The fertilized ovum divides over and over again, resulting in the organism, and so do the germs (the contents of the stirp). Dominant germs result in the various units of the organism. Residual germs from all these divisions are the sexual elements. Galton provided little clarification or details about 'residual germs'. The implication of this view of the separation of sexual elements from other units of the body was that, contrary to Darwin's view, changes to those other units did not affect the germs in sperm and ovum. Hence acquired characteristics could not be inherited. Pangenesis was wrong.

These are the conceptual underpinnings of Galton's theory. He also gives it some mathematical substance. As noted, Darwin's knowledge of mathematics was limited. Moreover, he was suspicious of mathematical proofs that seemed at odds with empirical evidence. George Darwin, Charles Darwin's second son, however, excelled in mathematics at Cambridge. In 1868 he was second wrangler. Consequently Darwin turned to George for assistance when mathematical presentations taxed his ability. Based on his correspondence with George, Darwin wrote to Galton on 18 December 1875, agreeing that the gemmules on his pangenesis theory largely reproduce in the reproductive organs. His countermove was to raise an objection to Galton's view. He provided a phenomenon that he believed

Galton's view could not explain. Moreover it supported pangenesis. He wrote,

> If 2 plants are crossed, it often or rather generally happens that every part of the stem, leaf – even to the hairs – and flowers of the hybrid are intermediate in character; and this hybrid will produce by buds millions on millions of other buds all exactly reproducing the intermediate character. I cannot doubt that every unit of the hybrid is hybridised and sends forth hybridised gemmules.

On 19 December, Galton responded. The response is, as expected, mathematical. He points out to Darwin that according to his own view, hybridization is entirely explainable. Moreover, he extends the explanatory power of his theory:

> If there were two gemmules only, each of which might be either white or black, then in a large number of cases one-quarter would always be quite white, one-quarter quite black, and one-half would be grey. If there were three molecules, we should have 4 grades of colour (1 quite white, 3 light grey, 3 dark grey, 1 quite black and so on according to the successive lines of 'Pascal's triangle').

As his student Karl Pearson remarked, 'This letter shows how very closely Galton's thoughts at this time run on Mendelian lines.' Galton's reference to Pascal's triangle would have eluded Darwin (illus. 11).

The number in each row, after row 0, is the sum of the two numbers above it. So, in row 1, the number above each of the two units is 1; in row 2, the numbers above 2 are 1 and 1 (1+1 = 2); in row 3, the numbers above each of the 3s are 1 and 2 (1+2 = 3); in row 4, the numbers above the 4s are 1 and 3 (1+3 = 4) and above the 6 are 3 and 3 (3+3 = 6); and so it continues. Down the sides of the triangle are only 1s because only the single number 1 is above them.

Galton used Pascal's triangle to determine quickly and easily the number of grades of colour that would result from different hybridization situations. As he noted, 'If there were three molecules, we should

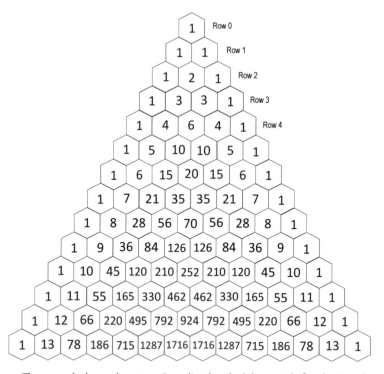

11 The numerical array known as Pascal's triangle: it is named after the French mathematician and philosopher Blaise Pascal but dates back to at least the middle of the 13th century in the writings of the Chinese mathematician Yang Hui and is found in Persian mathematical texts.

have 4 grades of colour (1 quite white, 3 light grey, 3 dark grey, 1 quite black and so on according to the successive lines of "Pascal's triangle")'. But this, as he suggests, can be extended to more than two molecules. This is where Pascal's triangle is useful. Hence, with three molecules, there will be four grades of colour. The ratios of these grades of colour will be 1:3:3:1 (row 3 of the triangle), which also can be expressed as fractions: $^{1}/_{8}$, $^{3}/_{8}$, $^{3}/_{8}$, $^{1}/_{8}$. If there were four molecules, there would be five grades of colour in the proportion given in row 5. Pascal's triangle allows the ratio/proportion of each grade of colour to be determined easily.

Ingenious though Galton's attempt to develop a theory of heredity was, he made an arguably greater contribution. Galton was interested in traits of organisms that vary quantitatively – height, weight, number of body hairs and speed of running, for example. These traits all have measurable differences, from few to many, from less to more and so on. They can be counted: inches of height, kilometre per second. The significance of his research was that he found that those traits that he investigated conformed to a normal distribution, best known today as the bell curve (see illus. 12). Connecting this to his theory of heredity, the germs in stirps were the basis of this variation of each trait. Traits, he postulated, are frequently the result of many germs working in concert and could be affected by the environment. Nutritional deficiencies could result in underdeveloped height, for example. In this he anticipated the contemporary field of quantitative genetics, a sophisticated extension of Mendelian genetics. Contemporary quantitative genetics takes into account that most traits – especially in mammals – result from the interaction of many genes, can be influenced by the environment and vary quantitatively in exactly the way Galton described. The bell-shaped line in the graph indicates the number of individuals with a particular value of a trait (illus. 12).

Height is an excellent example of something Galton discovered and described mathematically. According to a 2010 survey by the UK Health and Social Care Information Centre, height in male humans in England (age 25–34) clusters around 1.776 m (5 ft 10 in.);

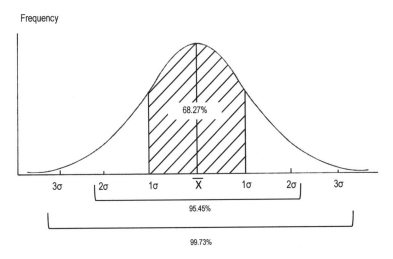

12 A normal (Gaussian) distribution. The tails – above and below 3σ – are very rare traits (135 organisms per 10,000; σ is known as standard deviation).

34.135% of males in England have a height between 5 ft 10 in. and 6 ft 0.8 in., and 34.135% have a height between 5 ft 8.2 in. and 5 ft 10 in. Since the values differ for men and women and for different ethnic groups, each group must be examined separately; for refined precision social class, partly due to nutritional differences, must be considered. So, for English males, 68.27% have a height between 5 ft 8.2 in. and 6 ft 0.8 in. Only 15.865% are shorter than 5 ft 8.2 in. and only 15.865% are taller than 6 ft 0.8 in. At the extremes, there will be only one person with that height. For example, the tallest man on record in 2012, Robert Wadlow, was 2.72 m (8 ft 7.7 in.). He was at the extreme upper end of the distribution and could be considered, in Darwin's sense, a 'sport' because he suffered from a hypertrophy of his pituitary gland. The shortest man on record, Chandra Bahadur Dangi, is 54.6 cm (1 ft 6.6 in.). He is at the extreme bottom end of the normal distribution. Extreme cases are usually the result of physiological abnormalities.

This work of Galton's is the foundation of biometrics: the measurement (metrics) of biological (bio) traits. Galton's work on traits such as height was exceptionally valuable. His work on intelligence as a trait was more controversial. Height is an observable trait. Intelligence is a more elusive concept. At the time, his views on intelligence were not particularly controversial. What is controversial today about his work is his use of biometrics, especially intelligence, in support of eugenics.

Galton and Darwin mostly approached heredity theoretically. August Weismann's work approached it empirically but he did have an important conceptual framework. Weismann staunchly defended natural selection as *the* evolutionary mechanism. On this, his view was even stronger than Darwin's. He differed from Darwin on two central points, however, rejecting the inheritance of acquired characteristics and the ability of the organs of the body to influence (contribute to) the reproductive substance. Weismann's conceptual framework distinguished sharply between reproductive material, which he called the germ plasm, and somatic material (from the Latin *soma*, meaning body: that is, cells of the body). The former persists over generations. It is more or less immortal and is passed to offspring through the ovum or sperm. He rejected the heritability of acquired characteristics because somatic cells are separate from and have no influence on the germ plasm (reproductive cells: sperm and ova). Hence any environmental impact on organs or cells of the body cannot be passed on through the germ plasm. The germ plasm contains everything necessary to create an organism, but the soma has no effect on the germ plasm. In addition to being nearly immortal, the germ plasm is passed to the next generation intact.

Weismann's early empirical support for this conceptual view came from experiments and microscope observations. He worked mainly on *Hydra*, a genus of hydrozoa. *Hydra* is a freshwater invertebrate that resembles a small thin tube somewhat longer than 1 inch (about 20–30 mm). It is closed at one end and open at the other. At the open end it takes in food and expels any waste.

Weismann established a lineage of germ cells and observed that during an organism's development from a single cell (the fertilized ovum) to the mature adult, the germ cells remained constant. On the basis of this, he postulated that (1) the germ cells remain constant during development from a single cell to the adult organism and (2) untransformed germ cells were transmitted to the next generation. He also observed that fertilized sea urchin eggs underwent uneven division, an observation that some have taken to be the beginning of the understanding of meiosis, especially in light of what Weismann called the process of reduction.

Somatic cells reproduce by duplicating the chromosomes in the nucleus, pinching the cell membrane in the centre and then splitting into two cells, each with one set of chromosomes; this is called mitosis. Each resulting cell has a complete set of homologous (matched) chromosomes. What today are known as gametes (sex cells) and what Weismann called germ plasm begin by undergoing a division similar to mitosis. Once there are two cells, however, each set of chromosomes separates into single strands and then division into two cells from each original occurs, resulting in four cells, each with only one strand of the original sets of chromosomes. This is cell meiosis (illus. 13).

The beginning of cytology is linked to the techniques for grinding lenses and building microscopes. As an aside, Galileo produced some of the finest early lenses, which he used to build his telescopes.

MEIOSIS

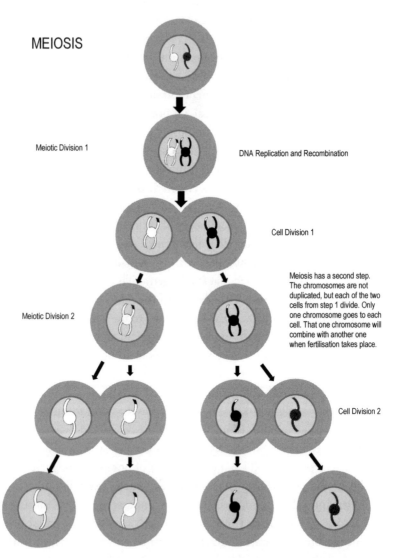

Meiotic Division 1

DNA Replication and Recombination

Cell Division 1

Meiosis has a second step. The chromosomes are not duplicated, but each of the two cells from step 1 divide. Only one chromosome goes to each cell. That one chromosome will combine with another one when fertilisation takes place.

Meiotic Division 2

Cell Division 2

13 Meiosis. There are two steps in this process. The first is the same as cell replication (mitosis): the chromosomes duplicate and the original cell divides into two, with one set of chromosomes going into each of the cells. In the second step, the chromosomes are not duplicated, but each of the two cells from step 1 divides. Only one chromosome goes to each cell; that one chromosome will combine with another when fertilization takes place.

He also invented the microscope, although his use of it was minor compared to his discoveries using the telescope. The biological use of the microscope dates back to the early part of the seventeenth century. The publication of Robert Hooke's *Micrographia* in 1665 marks the first major treatise of discoveries using it. In 1861 Franz von Leydig defined the cell as a clump of protoplasm containing a nucleus. In 1873 Anton Schneider described chromosomes and by 1883 their behaviour was understood. Hence a great deal had been discovered by 1893, when Weismann published his *The Germ Plasm: A Theory of Heredity*.

For Weismann heredity actually played a background role. Transmission from generation to generation had to occur for evolution to be possible but, in his view, nothing dynamical rested on that transmission. The germ plasm of an organism was nearly immortal and, hence, for each organism, there was a germ-plasm line from ancestors through to progeny. What shaped everything was natural selection. It was the only dynamical mechanism. It explained why organisms with certain traits were more prevalent. It explained how, over time, the traits of organisms changed and eventually new species arose. It explained degeneration, which was the result of the absence of selection on a trait. Unlike Galton's work, Weismann's views had little lasting impact on evolutionary theory. The exception is the concept of cell lines and the term 'germ plasm', which is still used but is not believed to have exactly the properties Weismann espoused. His views on selection as the only dynamical force in evolution fell by the wayside. His views on the separation of kind between somatic cells and germ plasm are still employed, though they have been modified. Gametes are no longer considered 'immortal'. In general, his ultra-selectionist view, which then was known as neo-Darwinism, impeded the acceptance of Darwinian evolution. Many were of the view that if Weismann's description of Darwinian evolution was accurate, they were not Darwinians.

By the end of the nineteenth century there were four dominant positions. First, there was Darwin's, which he eloquently put forward in *The Origin* and held until his death. The four relevant elements of his view were that natural selection acted on small variations and not macro-variants such as 'sports'; that pangenesis accounted for hereditary transmission; that acquired characteristics were heritable; and that natural selection tamed what would otherwise be a rampant increase in variability. Second, there was Galton's view. The relevant element here was a rejection of pangenesis in favour of a stirps view. This almost reached Mendel's theory by a biometry route. He also, anomalously compared to his closest supporters, rejected the theory that natural selection acting on small variations would result in evolution. Like Huxley, he accepted natural selection but believed it acted on macro-variations. Third, there was Weismann's view that natural selection was the sole cause of all aspects of evolution and that the units of heredity were the germ plasm, which was preserved from generation to generation and not affected by changes in, or to, the somatic cells of the body.

The fourth view has not yet been explored. The major character who advocated this view was William Bateson. Bateson's early view was that natural selection was a mechanism of evolutionary change but, like Thomas Huxley, he believed it could act only on organisms that were major deviations within a species – what had been referred to earlier as 'sports'. He rejected Darwin's view that it could act on small individual variations. Bateson had already fallen out with two followers of Galton: Weldon and Pearson. This happened before the end of the nineteenth century. At the dawn of the twentieth, things became extremely nasty between Bateson, Weldon and Pearson.

The intellectual to-ing and fro-ing in the period after *The Origin* regarding heredity, variation and the things on which selection acts is understandable. The history of science makes clear that in the early periods of all major scientific advances, debates about the fundamental assumptions of a theory abound. A classic example is the debates between the Newtonians, like Samuel Clark, and Gottfried Leibniz with respect to time and space. Clarke was a friend of Newton and defended Newton's view that space and time were absolute; that is, no matter where in the universe you are the passage of time is always the same, and all things that move move in a space that is fixed. Leibniz argued that time and space are relative: the passage of time is different for someone on a moving boat than it is for an observer on the shore, and space is defined completely by the object and its behaviour within space. After the acceptance of Einstein's Special

Theory of Relativity (time is relative) and General Theory of Relativity (there is no fixed absolute space), Leibniz more or less triumphed over Newton. The correspondence between Clarke and Leibniz is the most revealing of the clashes of views of Newton and Leibniz.

What separates the debates over evolution from those in seventeenth-century physics is the lack of a mathematical formulation of Darwin's theory. This meant that elements of the theory were informal and vague. Galton came very close both to Mendel's theory and a mathematization of evolution, but what he offered lacked evidence and, above all, clarity. As we shall see, Mendel provided both. Weismann's theory had some essential features of heredity at the cell level but had too definite a separation of somatic cells and germ cells.

Galton had come close to a conceptual version of Mendel's theory. One might therefore expect that debates about the hereditary basis of evolution would subside once Mendel's work came into prominence in the early twentieth century. Subside they did not, however. Indeed, the coming to prominence of Mendel's experiments and theory simply fuelled the feuding. It became more intense, personal and dysfunctional.

So, as we come to the close of the nineteenth century, the fact of evolution – that it had occurred and had given rise to an array of living forms – was widely accepted in Britain and Continental Europe. By 1900 Darwin's theory was widely known. *The Origin* had been translated into Danish, Dutch, French, German, Hungarian, Italian, Polish, Russian, Serbian, Spanish and Swedish. In short, it had been translated into almost all European languages. Today it has been translated into 29 languages, second only, for a scientific work, to the first books of Euclid.

In 1900 its influence and impact affected every aspect of biology. Its impact on religious thought was profound. The major denominations sought either to make compatible faith and evolution, or to demolish evolution. The latter task became more difficult with every new biological discovery. Political and social life was affected by things like the eugenics movement. This employed biometrics, and a questionable understanding of genetics and the mechanism of natural selection. It also rested on suspect values (the ideal and pure population, for example). The entire fabric of British and European science and society had been shaped by evolution and Darwin's theory. Resolving the four debates highlighted in this chapter became more pressing as evolution became more widely accepted. The nineteenth century ended without a resolution, although Galton had come very close.

Things were different in the United States. There are complex reasons for this. One important one was the total opposition to evolution of the influential Harvard University professor of palaeontology (Jean) Louis Agassiz. He ruled the Museum of Comparative Zoology at Harvard with an iron hand and was a deeply religious man committed to special creation. Another reason was that the u.s. was different in its religious history and structure. Deeply religious, persecuted and non-mainstream sects had settled there. It began with the pilgrims (the Brownist sect, which adopted the views of Richard Clyfton). While theology in Britain and Europe became more critical of the Bible as history and science, in the u.s. it remained literalist.

FOUR

Mendel Unlocks One of Nature's Secrets

Evolution by natural selection requires that traits be inherited. In the previous chapter we saw how different contributors to the debates about evolution viewed the mechanisms of heredity. Although the fact of heredity seemed unquestionable – offspring do resemble their parents – no adequate mechanism for it was known. To be more accurate, the overwhelming majority of scientists in the late nineteenth century did not know it. By 1865, one man did.

In 1854, 750 miles from Darwin's home in Down, Kent, an Augustinian monk began work on plant hybridization at St Thomas's monastery near Brno, Moravia (modern-day Czech Republic).

Gregor Mendel was born on 22 July 1822 in Heinzendorf, Austria (illus. 14). He came from a peasant farming background, which some have suggested explains his interest in plants. At the age of eleven he entered a secondary school in Troppau. The headmaster was an Augustinian monk in the monastery in Brno – the capital of Moravia. In 1840 he graduated with honours. He then attended the Philosophical Institute, part of the University of Olmütz. He graduated in 1843, again distinguishing himself in all subjects but philosophy. Although it was only a two-year programme, Mendel suffered from depression and had to take time away from his studies. His financially impoverished background always meant that pursuing his studies was a burden on his parents and himself. He excelled in physics at the institute and his physics professor, recognizing his poverty, recommended him as a novice to the Brno monastery. After one year he entered theological training. After completing his theological studies, Mendel was ordained in 1847. Shortly after, the

14 Gregor Mendel, 1822–1884.

abbot of St Thomas's agreed to allow him to study natural science and mathematics at the University of Vienna, which he did for two years.

Major scientific breakthroughs frequently depend on many factors falling into place. This was the case with Mendel. One factor was that he was intellectually gifted. That he excelled at his studies was clear evidence of this. Three other factors were also important. One was his training at the University of Vienna; the second was the fact that St Thomas's was steeped in science; and the third was that St Thomas's had an excellent experimental botanical garden and a herbarium. Few monasteries had this depth of interest in science. These all contributed to the possibility of his major scientific breakthrough. It was in part Mendel's astute choice of monastery and in part good fortune that landed him in a place that fostered his work. Another feature of landmark transformations in science is that it takes an insightful genius to make an important discovery. Once made, the leap often seems obvious to everyone. This was true in Mendel's case, although it took 35 years for his discovery to be widely known.

Unlike the theory of evolution, Mendel's experimental work and his findings were consistent with his religious views. None of it challenged the special creation of the cosmos by God. His research was revealing only the wonders of God's creation. There is evidence in Mendel's writings that he read *The Origin* but rejected the fact of evolution and Darwin's theory explaining it. He held firmly to special creation by God. He believed that blind chance could not explain the array of living things, and especially humans. Evolution, unlike the results of his research, was clearly contrary to special creation.

In a twist of fate, Mendel's work at St Thomas's unlocked one of the mechanisms of heredity. The theory he constructed was exactly what Darwin's theory lacked. It was vastly superior to the theories that Darwin, and even Galton, had postulated. In a variety of ways their theories were incomplete and controversial, and lacked empirical support. Mendel's paper was published in the 1865 *Proceedings of the Natural History Society of Brünn*, which appeared in 1866, only seven years after the appearance of the first edition of *The Origin* and three years before the fifth edition of 1869. Mendel sent copies of his paper to many universities, other institutions and individuals, but his work largely went unnoticed. The paper was listed in the Royal Society of London's 'Catalogue of Scientific Papers' for 1866. Despite this, there is no evidence that Darwin, or his scientific colleagues in Britain, read the paper, though there has long been a rumour that a copy of the *Proceedings* containing Mendel's paper was found in Darwin's library after his death. According to the rumour, the pages were uncut (many publications at this time were printed and bound with alternate pages folded on the right-hand side and readers had to cut the pages in order to read). There is no evidence that this rumour has a basis in fact. It does, however, highlight the fact that the mechanism of heredity was at hand for Darwin but went unnoticed.

The essence of Mendel's groundbreaking experiments and the theory he constructed to explain the results are easy to grasp. Given the importance of his theory to evolutionary theory, it is worth looking at it in some detail.

Understanding Mendel's Experiments

Mendel was interested in hybridization in plants (interfertilizing two varieties of a plant) and set out to discover what happens over many generations. First, he had to select a species of plant with the ideal characteristics. Then he had to create hybrids and then breed them. The details of this are set out below.

Mendel was explicit about his goal; he was seeking to discover generally applicable laws. His success was in part due to his commendable attention to experimental details. He kept detailed records of the number of distinct forms, the organization of progeny by generations and the statistical (numerical) relations of progeny in each generation. He designed the experimental conditions and steps in a way that maximized the probability that the results would shed light on the general laws of hybridization. His execution of the experiments was exceptional.

The Seven Key Steps in the Experiment

1) Selecting the Ideal Plant
The plants must be selected carefully, so that the data obtained from the experiments are unambiguous. Mendel concluded that this meant:

- They must have clearly identifiable characteristics (colour, shape, and so on);

- The offspring in each generation must be able to be protected from contamination by foreign pollen;

- They must remain equally fertile in all subsequent generations; and

- The original plants (the initial generation) must have bred true through many generations; that is, all the offspring of each type of plant must generation- after-generation have exactly the same traits as the original parents of that type (Mendel took nothing for granted, so he cross-fertilised within type for several generations until he was satisfied that he had pure-bred types).

Mendel chose peas of the genus *Pisum* because they met these requirements.

2) Careful Observation

Observation of successive generations must be carefully undertaken to ensure that each generation is identified and kept separate from other generations, and the results obtained must be accurately assigned to the correct generation. Designators for different generations simplify this task. An F with a numerical subscript designates generations. F_0 designates the initial generation. In Mendel's experiments, there were different types of parents each with distinguishable character- istics: seed colour, seedpod colour, seed size, seed texture and so on. The ones on which he focused are listed in 3) below.

F_1 designates the generation that results from breeding one type of parent with the other type, producing a hybrid plant. The traits of the plants of this generation vary. F_2 designates the generation resulting from the interbreeding of the hybrids, and so on.

3) Selecting Parent Seeds (F_0) and the Characteristics to be Studied

Mendel obtained seeds of 34 more or less distinct varieties, which he checked to ensure that they bred true; that is, in each generation exactly the same characteristics were present. He then selected 22 varieties for the experiment. He focused on seven characteristics, which differed in each parental variety:

- Round vs wrinkled peas

- Yellow vs orange peas (seen through the transparent seed coats)

- Seed coats white vs grey, grey-brown, leather brown

- Smooth vs wrinkled ripe seed pods

- Green vs yellow unripe seed pods

- Axial vs terminal flowers

- Long vs short stems (he chose 6–7 ft and 0.75–1.5 ft),

4) Creating Hybrids
He crossed numerous F_0 plants with each character difference to obtain hybrids (F_1 generation).

5) Breeding Hybrids
Mendel intrafertilized the hybrids to obtain the F_2 generation. This is an important step because it allowed him to begin his investigation of the proportion of each trait that arises from the crossing of hybrids. Those proportions, he hoped, and later was to confirm, would yield important information about the underlying structure (later known as the genetics) of hybrids (illus. 15).

6) The Next Generations ($F_2–F_4$)
Mendel then intrafertilized plants from F_2 with the same characteristics. He fertilized plants that all produced round seeds with pollen from plants that also all produced round seeds, and he fertilized plants that all produced wrinkled seeds with pollen from plants that also all produced wrinkled seeds.

Pollen

	(R)	(W)
Ovum (R)	(RR)	(WR)
(W)	(RW)	(WW)

15 The outcomes of crossing hybrid plants. All the hybrid plants have the factor-pair *RW*. Offspring do not.

7) Recording the Number of Plants with Each Characteristic within Each Generation

Mendel kept careful records of the number of plants with each characteristic within each generation. He then looked at the proportions of each characteristic in each generation.

The Observed Ratios

F_1 generation
Mendel found that in the F_1 generation (step 4) only one characteristic of each pair of characteristics (round vs wrinkled peas, for example) was present in all the progeny.

F_2 generation
Mendel expressed his result as ratios. (As explained previously, : is the symbol for ratio. $A:B = 4:1$ is read as 'the ratio of As to Bs is 4 to 1', meaning that for every five things, 4 will be A and 1 will be B. This could be expressed as 80 per cent are As and 20 per cent are Bs or as $4/5$ are As and $1/5$ are Bs. Ratios are easier to use for the kinds of calculations and comparisons Mendel wanted to employ.)

In the F_2 generation (step 5) he observed, that for each pair of characteristics in F_0, they emerged in F_2 in the ratio 3:1 (three plants producing round peas: one wrinkled, for example).

F_3 Generation:
In the F_3 generation (step 6) Mendel observed that, when plants that produced wrinkled peas (W) in the F_2 generation were intrafertilized, they bred true to form in all subsequent generations; that is, all the plants produced W-peas in the F_3 generation and in all the generations thereafter. Plants that produced round peas (R), when intrafertilized, yielded R:W = 3:1 (illus. 16).

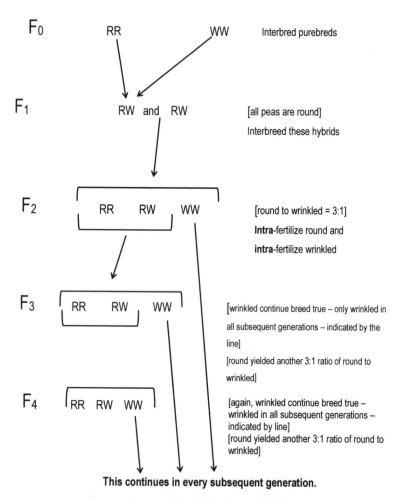

F₀ RR   WW   Interbred purebreds

F₁ RW and RW   [all peas are round]
Interbreed these hybrids

F₂ RR   RW   WW   [round to wrinkled = 3:1]
Intra-fertilize round and
intra-fertilize wrinkled

F₃ RR   RW   WW   [wrinkled continue breed true – only wrinkled in
all subsequent generations – indicated by the
line]
[round yielded another 3:1 ratio of round to
wrinkled]

F₄ RR RW WW   [again, wrinkled continue breed true –
wrinkled in all subsequent generations –
indicated by line]
[round yielded another 3:1 ratio of round to
wrinkled]

This continues in every subsequent generation.

16 Mendel's breeding pattern. The right-hand side illustrates that Mendel
always observed the wrinkled-pea characteristic when wrinkled-pea plants were
bred with wrinkled-pea plants. The left-hand side illustrates that he observed
the 3:1 ratio when round-pea plants were bred with round-pea plants.

Mendel's Explanation of these Results

Mendel's genius is manifest in two ways. The first is his ability to design and conduct experiments that yielded clear results, as demonstrated above. The second is his power to construct a mathematical theory to explain these results. There are several essential components to a mathematical theory. The goal of a theory is to specify a mechanism that explains a class of observations. In this case, what needs to be explained is why when plants with seeds manifesting one of the two characteristics (smooth, for example) are intrabred, the seeds of the plants in the next generation occur in a 3:1 ratio (smooth to wrinkled, for example), whereas when seeds from plants with the other characteristic (wrinkled) are intrabred, they produce only plants with this other characteristic.

An explanatory mechanism requires objects. The characteristics of these objects, along with a description of how they behave, constitute the mechanism. The objects in Mendel's theory are 'factors'. These were not visible at the time he did his experiments. Hence he was positing that they exist. His factors have three central characteristics. First they normally exist in pairs. Second, different pairs yield different characteristics: round peas, wrinkled peas, yellow peas and orange peas, for example. Third, some are dominant and others recessive. That is to say, when the factors in a pair are different, one will be expressed rather than the other.

Mendel now specifies how these factors behave over time. The pairs segregate (become unpaired), with one of each pair going into pollen, the other into the ovum. When an ovum is pollinated (fertilized) the factors recombine. The recombination is random. Now imagine, as Mendel did, that round or wrinkled peas are caused by two factors: R and W. There are four possible pairings: RR, WW, RW and WR. Mendel assumed that RW and WR caused the same trait in the plant, so that their effect on the plants was identical. Mendel also assumed that RR causes round peas and WW causes wrinkled peas. These are the plants that breed true for generation after generation when interbred. When he crossed these true-breeding plants, he assumed that each offspring would have one R and one W. This results from the factors RR segregating, with one R going into the pollen and the other into the ovum. The same is the case for the factor W. When the pollen from a plant that produces round peas is used to fertilize the ovum of a plant that produces wrinkled peas, the pair of factors in that fertilized ovum will be RW. Indeed, all such crossings will yield plants with the factors RW. As we know, when Mendel did this, all the plants produced round peas. That is the effect of dominance; R dominates over W.

This model allowed Mendel to deduce the ratios of traits in each generation. Assume in F_0 generation that all the peas yield either RR or WW plants. In this case, Mendel claimed, the pollen and ovum of the RR plants will all contain R factors and the pollen and ovum of WW will all contain W. If ova from RR plants are only fertilized by pollen from WW plants and vice versa, all the pairs of factors in the first generation will be RW. Since R is dominant, all the peas produced by the plants in this F_1 generation will be round, which is what Mendel observed.

Next, he cross-fertilized these RW plants – that is, he used the pollen of one plant to fertilize the ovum of another plant. He made sure that none of the plants self-fertilized; they all cross-bred. The pollen contains either an R factor or a W factor; the same is true of the ovum. As a result, some R-factor pollen will fertilize an R-factor ovum; some R-factor pollen will fertilize a W-factor ovum (and vice versa); some W-factor pollen will fertilize a W-factor pollen. The result will be some RR plants, some WW plants and some RW plants.

As the figure on page 84 makes clear, in the F_2 generation, a quarter of the plants will be RR, a quarter will be WW and half will be RW/WR. Since R is dominant, RW plants will produce round seeds. Consequently a quarter of the peas will be wrinkled (the WW plants) and three-quarters will be round (one-quarter RR + one-half RW/WR). Mendel's model also predicts that if the quarter of the plants that produce wrinkled peas are bred, they will continue in each subsequent generation to produce plants that yield only wrinkled peas.

That is the essence of his mathematical theory. What he did was to use this theory to predict the ratios of round to wrinkled peas (and the other characteristics) in each generation. The fact that Mendel's actual experimental results were consistent with the prediction confirms the theory.

The confirmation of the theory was strengthened when he continued by examining multiple characters through the generations. The results revealed that each character conformed to the 3:1 ratio, as predicted mathematically by the theory. Mendel summed up his findings and his theory in this way:

> So far as experience goes, we find it in every case confirmed that constant progeny can only be formed when the egg cells and the fertilising pollen are of like character, so that both are provided with the material for creating quite similar individuals, as is the case with normal fertilisation of pure species. We must therefore regard it as certain that exactly similar factors must be at work also in the production of the constant forms in the hybrid. Since the various constant forms are produced in *one* plant, or even in *one* flower of a plant, the conclusion appears logical that in the ovaries of the hybrids there are formed as many sorts of eggs cells, and in the anthers as many sorts of pollen cells, as there are possible combination forms, and that the egg and pollen cells agree in their internal composition with those separate forms.

It is interesting to note in passing that having constructed his model using mathematics, Mendel joined a tradition reaching back to at least the English scholar Thomas Bradwardine, who, around 1326, wrote in his *Tractatus de Continuo*:

> It is [mathematics] which reveals every genuine truth, for it knows every hidden secret, and bears the key to every subtlety of letters; whoever, then, has the effrontery to study physics while neglecting mathematics, should know from the start that he will never make his entry through the portals of wisdom.

Almost 300 years later, in 1623, Galileo wrote the better-known version of this in *Il Saggiatori* (*The Assayer*):

> Philosophy is written in this grand book, the universe, which stands continually open to our gaze. But the book cannot be

understood unless one first learns to comprehend the language and read the letters in which it is composed. It is written in the language of mathematics, and its characters are triangles, circles and other geometric figures without which it is humanly impossible to understand a single word of it; without these, one wanders about in a dark labyrinth.

It reached its pinnacle, prior to the twentieth century, with Newton's mechanics, for which he had to develop the infinitesimal calculus. Although Gottfried Leibniz simultaneously developed the infinitesimal calculus and it is his notation that is used today, it was Newton who used it as the language in which to express his mechanics and deduce empirical consequences.

Darwin, as noted, stands as an exception in this tradition: his formulation of the theory of evolution did not employ mathematics as its language and there is not a single equation in *The Origin of Species*. That step had to wait until the early part of the twentieth century, after Mendel's work came into prominence. The first extension of Mendel's mathematical model came in 1905 with the mathematician G. H. Hardy and further development occurred in the 1920s with the mathematician Ronald A. Fisher and the geneticists J.B.S. Haldane and Sewall Wright. Because Darwin did not express his theory in mathematical terms, his theory was informal. The seven decades between the publication of *The Origin* in 1859 and the publication of Ronald Fisher's *Genetical Theory of Natural Selection* in 1930 were tumultuous in significant part because his exposition was informal. The next chapter follows the controversies and deep divisions that occurred among evolutionary biologists during those decades.

FIVE

Troubled Waters

As noted, Britain was changing rapidly in the period before and just after the publication of *The Origin*. Many of the important changes were scientific, technological and theological. In the last two decades of the nineteenth century and into the first decade of the twentieth, Britain underwent considerable social change. The horrors of the working conditions in the factories of the Industrial Revolution were abating; the air was getting cleaner and a discernible and growing middle class was emerging. Britain was moving into a new phase of democracy. By 1918, all males over the age of 21 were given the vote, as well as women over 30. Hence, 75 out of every 100 adults now had the right to vote. Women could also sit in the House of Commons as MPs. Trade within the empire was booming, although questions about imperialism were increasing and Britain's hold on the empire was beginning to be tenuous. Early in this period, Darwin died (on 12 February 1882). In 1901 Queen Victoria died. An era had ended, and with it the intellectual climate within the development of evolution was being debated. With a rising middle class, education became more widespread and science education became more prevalent. Many of the characters in the debates of the period covered in this chapter might never have risen to scientific prominence 50 years earlier. This different mix of individuals led to a different kind of scientific climate.

In this rapidly changing world, Weldon, Galton and Pearson had come together. First, we look at Weldon. His path from his days at St John's to this point in the story was different from Bateson's. A decade after those university days, when they worked together and were friends, he became interested in variation. As a morphologist, he had observed changes in larval and adult characters of the lizard *Lacerta muralis*. It appeared that evolutionary changes in the adult were accompanied by changes in the larval stage. Weldon, however, had no hypothesis about this connection between the changes in the adult and the larvae. He felt there was a connection but did not have available the conceptual tools to investigate and demonstrate it. In 1889 he read Galton's *Natural Inheritance*, as did Karl Pearson. Both were impressed with the quantitative methods Galton had developed. He began measurement on several organisms and found that variations were distributed in the way Galton had found for humans. This work was the beginning of Weldon's biometric approach but he did not have the mathematical competence to provide more than a rudimentary analysis. Although he tried to learn some probability, given the complexity of the task he was never entirely competent or comfortable. Fortunately, his work came to the attention of Karl Pearson who, as indicated, was also enamoured of Galton's work; Pearson did have the mathematical skills required.

By 1894 Weldon, Galton and Pearson were an intellectual team as well as being friends. Pearson and Weldon initially formed a friend-ship when they teamed up to oppose a plan to unite the two founding colleges of the University of London, King's College and University College, into a new university, Albert University. At the time, the two colleges were the teaching centres of the university, and the uni-versity itself was mainly responsible for examinations. Weldon threw himself into the fight: he spent an evening collecting signatures, from academic staff in the university, on a petition opposing the cre-ation of the new university. On next day the petition was delivered to all the Members of Parliament. Although Queen Victoria was not amused, the bill creating the Albert University by giving it a charter was abandoned. That victory was only the first step. Weldon, Pearson and some other academics created the Association for Promoting a Professional University of London. The various colleges, at the time, were almost completely independent and linked only by a loose fed-eration agreement. This new association, of which Thomas Huxley agreed to be president, advocated for a more unified university. Although both Weldon and Huxley, surprisingly giving his early fiery defence of evolution, were compromisers, Pearson was not. When the whiff of compromise was in the air, Pearson resigned as secretary of the association and Weldon took his place. Pearson continued to press for a more unified university, writing to *The Times* and moving amendments to Huxley's presentation of a vision to the executive committee. They remained close friends, having lunch together before lectures and debating the major issues on which they were working as well as brainstorming new ones. Pearson had initially rejected Galton's *Natural Inheritance*; Weldon convinced him otherwise. At that point, they met and corresponded with Galton frequently.

The three founded the journal *Biometrika* in October 1901, and Weldon and Pearson were the mathematical minds, especially Pearson, behind Galton's quantitative approach, which came to be known as biometry. Both espoused this approach with vigour. Nonetheless, they did not agree on everything. On the important issue of continuous versus discontinuous evolution, their views were in opposition. Galton had long held that evolution was discontinuous. Pearson, although a strong advocate of Galton's other views, had concluded in the 1880s, contrary to Galton, that evolution was continuous and held to this view throughout. Weldon concurred.

By 1890 Bateson had rejected completely the Darwinian view of natural selection acting on small individual variations and leading to continuous evolution – an insensible transition from one form to another. Bateson embraced and began to champion discontinuous (saltation-based) evolution. In a paper of 1891 he made clear his break with Darwinism. A break with Darwinism entailed a break with Weldon and Karl Pearson. Pearson and Weldon met the same year at University College, London; they quickly became collaborators. Both were committed Darwinians. Hence, Bateson was now offside.

This difference regarding natural selection came into stark relief with the publication in 1894 of Bateson's *Materials for the Study of Variation, Treated with Especial Regard to Discontinuity in the Origin of Species*. Weldon's review was very critical but respectful. Bateson, on the other hand, had been confrontational for some time. That confrontational stance entered a public phase on 28 February 1895 at a meeting of the Royal Society; it then continued. Most of it appeared in the pages of the prestigious journal *Nature* in the form of letters. After about three months, in June 1895, the editor of *Nature* refused to publish any further exchanges. During this period, the break, personally, intellectually and civilly, between Bateson and Weldon became complete. Pearson had already written Bateson off as incompetent, a stance that hardened as Bateson's mathematical weakness became ever clearer and his break with Weldon solidified.

Galton, ever the gracious diplomat, was straddling the two camps. One was the biometricians/Darwinians, led by Weldon and Pearson, who adopted Galton's biometric method and accepted continuous variation and continuous evolution, the latter being consistent with Darwin's view but contrary to Galton's. The other camp, led by Bateson, rejected Galton's biometric method but adopted his commitment to discontinuous evolution. This divide and Galton's attempt to straddle it became more complicated in 1900, when knowledge of Mendel's work entered mainstream biology.

Mendel tried to ensure that his work reached major scientists; nonetheless, little attention was paid to it before 1900, when Hugo de Vries and Karl Correns each reproduced Mendel's results but did not appear to have read Mendel's paper. Indeed, it seems that they only fully appreciated the deep significance of their results when De Vries read Mendel. In his 1894 article 'About Half-Galton Curves as a Sign of Discontinuous Variation' De Vries makes no mention of Mendel. Instead, he draws on work done by Galton and provides Galton's results (1:2:1 ratios, explained in chapter Three). He saw his work as supporting Galton's use of Pascal's triangle to generate the same ratios. De Vries also describes dominance without using Mendel's terms or specific definition. Given the terminology in this paper and the case he develops, along with what he wrote in his 1900 article, 'On the Law of Disjunction of Hybrids', it is reasonable to believe that he arrived independently at Mendel's ratios and, drawing on Galton, came very close to Mendel's model. Mendel's priority, however, was not questioned. Important to De Vries was his view that Mendel's work supported his own 'law of pangene hybridization' as part of his pangenesis theory of heredity; his continuing commitment to a version of pangenesis, even after realizing how similar his findings were to those of Mendel, suggests that he still did not appreciate the full impact of Mendel's theory.

As we have seen, the importance of Mendel's work and theory centred on three things. First, he provided results from a number of well-designed, expertly conducted and meticulously recorded experiments on hybridization. Second, he provided a mathematical description of the results in terms of ratios. Third, he provided a theory to account for the ratios. That theory postulated objects ('factors' – today known as alleles), characteristics of these entities (they can be dominant or recessive, for example) and a description of how the objects behave over time. These entities give rise, as pairs, to the characteristics of the organism; they segregate in the reproductive organs (the reduction Galton identified as necessary and Weismann had observed in a somewhat different context); and they recombine in fertilization to create new pairs that determine the characteristics of the offspring.

As also noted, when Mendel constructed his theory, these factors were unobservable. The evidence for their existence, properties and behaviour was indirect. This, however, is true of the entities postulated by other theories; electrons, mu mesons (muons) and so on are not observable but the physical theory that postulates their existence is highly probable. The probability that Mendel's factors exist rested entirely on the probability that his theory was true. Hence Mendel's proof of his theory rests solely on his experiments. He deduced (predicted) from his theory certain outcomes. These deduced outcomes matched the results of his experiments. This agreement of the empirical results with what the theory predicts suggests that it is correct and consequently that the factors it postulates must exist and must have the properties and behaviours he stipulated. The fact that the empirical results are consistent with what the theory predicts does not prove conclusively that the theory is correct, but it does make its correctness probable. This is a matter of logic. A brief examination of this logical point will make clear why the theories we have so far examined, and those we will examine in subsequent chapters, are only probably true – a fact that is true of all theories.

Mathematical proofs are deductive. If the initial statements – the premises – of a mathematical proof are true, the conclusions drawn using the rules of inference *must* be true: their truth is necessary. Scientific proof is different: it is inductive rather than deductive. In this case, even if the premises are true, the conclusion *could* be false.

Consider a classic case of deduction:

A) If it is raining and there is no protective covering, the pavement will be wet.

B) It is raining and there is no protective cover.

These are the premises. From these one can conclude *with certainty* that:

C) The pavement will be wet.

(A) and (B) cannot both be true and (C) false: (C) necessarily follows from (A) and (B). This specific pattern of deduction, know as a syllogism, goes back to Aristotle and is known as 'affirming the antecedent'. In an 'If X then Y' claim, X is the antecedent and Y is the consequent. In the example above (B) is the same as the antecedent of (A), the part before 'then'. Mathematical proofs today are vastly more complex but the same principle holds: if the premises are true, the things concluded from them using deductive rules must be true.

Science involves a different process of reasoning. A simplistic way of characterizing this reasoning is based on a generalization from instances. We observe many instances of things that are thrown and we observe that they all eventually come to rest; we conclude from these observations that all things that are thrown eventually come to rest. As the philosopher David Hume taught, however, although a million instances might generate great confidence in the connection between events, the next instance could always be different. His point was that we observe events only. We 'decide' that they are connected but we do not 'observe' a connection. The more times event A is followed by event B, the more confident we are that the first is causing the second. Nonetheless, it might be a fluke that they have always been connected. Hence, the next time event A happens, it might not be followed by event B. In Hume's words, we cannot observe a *necessary* connection between the events. All we observe is the constant conjunction of the events and their consistent temporal ordering. The decision to assume a 'general rule' about their connection is a leap in reasoning. Hence, cause and effect are a leap in reasoning. There is nothing necessary about the reasoning from many instances to a general rule. The more sophisticated way of characterizing induction in scientific reasoning is that, first, a hypothesis (an idea that can be tested) is formulated; second, an experiment to test the hypothesis is devised; and, third, the experiment is conducted (a reasoning process known as the hypothetical-deductive method). Do not be fooled by the use of the word 'deductive' in this label. The deduction is real but it is from the hypothesis to the experiment. One deduces that if the hypothesis is true, then a certain empirical consequence will occur under certain conditions. Once this deduction from the hypothesis is made, induction enters. The structure is:

D) If the hypothesis is true, then empirical consequence Ω will be found under Ψ conditions

E) Empirical consequence Ω has been found under Ψ conditions

F) Therefore the hypothesis is true

This is, however, fallacious reasoning; it can lead to false conclusions. Consider again the rain example:

A) If it is raining and there is no protective covering, the pavement will be wet.

G) The pavement is wet.

H) Therefore it is raining.

This is a fallacy of reasoning known as 'affirming the consequent'. The pavement might be wet because I have my sprinkler on and it is spraying onto it, or I might be deliberately hosing it down to clean it. Hence (A) is true and (G) is true but (H) is not a necessary consequence. There are other ways in which the pavement could become wet, which means (H) could be false even though (A) and (G) are true.

Returning to hypothesis testing, once a hypothesis has been formulated, an experimental test can be deduced from it. That is, if Mendel's theory, which constitutes a complex hypothesis, is true, a number of empirical consequences can be predicted; experiments, such as crossing hybrids, can be devised to see whether a predicted consequence occurs: the predicted ratio of traits in this case. The logic of this reasoning is:

I) If Mendel's model is true, then if hybrids are crossed, the ratio of smooth to wrinkled peas will be 3:1 in the next generation.

J) Hybrids were crossed and the ratio of smooth to wrinkled peas was 3:1 in the next generation.

K) Therefore, Mendel's theory is true.

It can immediately be seen that this is an instance of the 'fallacy of affirming the consequent' as in the rain example above (A and G, therefore H). In the case of Mendel's theory, (I) and (J) could be true even if (I) is false. Hence there is no *certainty* in hypothetical-deductive logic. As the number of accurate predictions made using the hypothesis increase, confidence in it increases; nonetheless, certainty is never achieved. This highlights the difference between mathematics and science: mathematics employs deduction, whereas science employs induction. Scientific knowledge, as a result, can only ever be 'probable', never certain. What Mendel provided was a theory that, given his experimental results, appeared probable.

Darwin, using Whewell's methods, also employed the hypothetical-deductive method but he went further and employed another test of certainty. He demonstrated that his theory predicted and explained phenomena in a number of diverse domains: Whewell's consilience of inductions. Although Mendel had constructed a theory from which deductions could be made and those deductions were consonant with experimental results, it applied only to a specific phenomenon – hybridization – and only in plants. Indeed, only in a specific plant. He clearly believed that the theory applied to all sexually reproducing living things but there was no consilience of inductions; it explained the phenomena it was constructed to explain, nothing more.

As the twentieth century opened, this was about to change. In a single sentence at the end of an article published in 1902, Walter Sutton, a graduate student at Columbia University, mused:

> I may finally call attention to the probability that the association of paternal and maternal chromosomes in pairs and their subsequent separation during the reducing division as indicated above may constitute the physical basis of the Mendelian law of heredity. To this subject I hope soon to return in another place.

Return to it he did in 1903, when he provided a much more detailed account. In 1902 this was a bold and speculative hypothesis but by 1910 the hypothesis had received considerable experimental and theoretical support. Understanding the significance of Sutton's and other later observations, and how Mendel's theory explains the observed behaviour of chromosomes, extends the explanatory scope of Mendel's theory and achieves a consilience of induction. Although not yet a rich consilience, the first step has been taken. With Sutton's observation, Mendel's factors, in part, become observable. They are still not strictly observable but the behaviour of cells that produces gametes – a process called reduction in these early days and today called meiosis – is exactly what one expects given his theory. Chromosomes segregate and recombine in the same way as Mendel's factors. So although chromosomes are not his factors, it is a small leap to the conclusion that his factors are parts of chromosomes. Sutton's observation also brought the fascinating world of cytology, the study of cells, into the evolutionary domain.

Microscopy had been advancing quickly since about 1840 and by 1900 lenses had been developed that, along with advances in staining tissues, allowed impressive observations of the composition of cells and their behaviour. Cytologists saw under their microscopes hitherto unknown constituents of cells. Chromosomes were among them and they displayed a fascinating behaviour. In cells there were long strands of material enclosed in a membrane; different kinds of organisms had different numbers of these strands. The collection of these strands and fluid material, along with the membrane that surrounded them, was named a 'nucleus'. Some cells, those of bacteria for example, contained chromosomes but they were not enclosed in a membrane, meaning they had no nucleus. Most plant and animal cells, however, have a nucleus containing its chromosomes. The cells of these plants and animals, and the plants and animals themselves, are called eukaryotes, from the Greek *eu*, good, and *karyon*, meaning kernel or seed. Those lacking a nucleus are called prokaryotes.

Cytologists observed that cells underwent division: one cell divided into two new and identical cells. The creation of these two identical cells first required that a new pair of chromosomes identical to the original pair be formed. The original strands, which were tightly connected, giving the appearance of a single strand separated into two thinner strands, were observed using a microscope. Each of these strands then appeared to thicken again. Each thickened strand migrated to one end of the cell. The cell membrane pinched inward in the middle and separated into two cells. This process was called mitosis and resulted in two identical cells. The thickening that was observed, we now know, results from the creation of a complementary strand; the single strands after the initial separation became coupled double strands again. Consequently, the cell now contained two nuclei, each with identical strands. When the cell squeezes together at its centre and two separate cells appear, each has one of the nuclei. This is how cells reproduce and multiply.

Sometimes, the divided strands – the pairs – duplicate but do not join. Instead, the cell divides twice to create four cells, each with only half of the original joined strand. These cells from the double division are gametes. In animals they are the sperm and ova; in plants, pollen and ova.

From the vantage point of the present, one would predict that once Mendel's theory was known and especially after the cytological observations were reported, Darwin's theory would be strengthened. A robust theory of heredity had been handed to the Darwinians. Clearly Mendel's theory challenged Darwin's pangenesis theory of heredity and his acceptance of blending inheritance, but it is impossible to know how Darwin would have responded if he had read and understood Mendel. On the one hand, Galton had developed a theory that was remarkably close to Mendel's theory and Darwin did not adopt it, which suggests he would not have found Mendel's theory congenial to his evolutionary theory, at least not initially. Moreover, Darwin's mathematical skills were slight. Consequently, if he were faced with Bateson's strident defence of his interpretation of Mendel, one inconsistent with continuous evolution and other elements of Darwin's theory, Darwin would probably have done what Weldon and Pearson did. They rejected Mendelism because of Bateson's interpretation of its implication for evolution. On the other hand, if he had developed an association with someone like Yule, who thought Darwinism and Mendelism were compatible, he might have adopted Mendel's theory. Mendel's theory and continuous evolution, as it turns out, were compatible and Pearson came very close to demonstrating this; Bateson's interpretation of it was simply wrong. Such a demonstration would probably have won Darwin over.

Mendel's theory provided Bateson, or so he thought, with exactly what he needed in his combat with the biometricians. Mendel's discrete factors were, according to Bateson, inconsistent with Darwin's blending inheritance. That it was incompatible with pangenesis and blending inheritance was clear, but whether this was a problem for Darwin's theory was less clear. As a result of Bateson's interpretation of the implications of Mendel's theory for Darwinian evolution, the theory failed to strengthen Darwinian evolution in the first decade of the twentieth century; if anything, it weakened it.

This controversy over which general view was correct soon reached an acrimonious stage. As will now be clear, on one side, Bateson championed Mendel's theory and claimed that Darwinian natural selection (gradualism) and continuous evolution were inconsistent with it. On the other side, Karl Pearson and Weldon embraced the biometric method of Galton and claimed that Darwin was correct about natural selection as the mechanism of evolution and that it acted on small individual variations; the continual production of variation was the real challenge yet to be solved. Just how poisoned the atmosphere had become by 1902 can be seen in the statistician G. Udny Yule's comments in *The New Phytologist* on Bateson and Saunders's

Report to the Evolution Committee and Bateson's *Mendel's Principles of Heredity: A Defence*. Edith Rebecca Saunders (1865–1945, referenced in published works as Miss E. R. Saunders) was a geneticist and plant anatomist. She collaborated with Bateson on heredity. The report on which Yule commented was one presented to the Evolution Committee, a recently formed special committee of the Royal Society.

> The sections of the two volumes which do appear to call for criticism and review are those relating to the bearing of Mendel's results on the conceptions of heredity in general, and on the work of Mr Francis Galton and Professor Pearson in particular. Mr Bateson devotes many words to these questions, but one cannot help feeling that his speculations would have had more value had he kept his emotions under better control; the style and method of the religious revivalist are ill-suited to scientific controversy. It is difficult to speak with patience either of the turgid and bombastic preface to 'Mendel's Principles,' with its reference to Scribes and Pharisees, and its Carlylean inversions of sentence, or of the grossly and gratuitously offensive reply to Professor Weldon and almost equally offensive adulation of Mr Galton and Professor Pearson. A writer who indulges himself in displays of this kind loses his right to be treated either as an impartial critic or a sober speculator. Mr Bateson is welcome to dissent from Professor Weldon's opinions, but it would have been well if he had imitated the studied moderation and courtesy of his [Weldon's] article.

The feuding continued until Weldon's death in 1906, after which Pearson moved on to other topics. Although the acrimony at a personal level diminished, the divide between Mendelism and biometrics/Darwinism continued. It was in this environment that R. C. Punnett, a committed Mendelian, gave a lecture to the epidemiological section of the Royal Society of Medicine on 28 February 1908 titled 'Mendelism in Relation to Disease'. During the discussion that followed, Yule expressed that he had less optimism that Mendelism would yield the benefits in medicine that Punnett had claimed. Yule, in his comment, was not rejecting Mendelism, although a few commentators have cast his remarks that way; as already noted, he was in fact one of the few who held that Mendelism and biometrics were entirely compatible. Most of his objections to Punnett's paper were focused on whether, in this or that case, more could be gained by a biometric (accounting) approach to a disease than a Mendelian one, even assuming Mendel to be correct with respect to the germinal description. This can be seen clearly in the published version of his comment:

> After all, what had to be dealt with was the character which was exhibited, and he agreed with the last speaker that in cases like these the actuarial method was likely to yield more valuable information to the medical man than a discussion on the basis of germinal laws, which might hold for the germ-cells but need not hold for the body, seeing how much the element of circumstance entered into the matter. Other factors as important as heredity must be taken into consideration. The actuarial statement included what the germinal statement did not, namely, those factors of disturbance which were of equal importance with the factors of pure heredity.

In the course of his comment, however, Yule asserted:

> The same applied to the examples of brachydactyly [abnormally short fingers and toes]. The author said that brachydactyly was dominant. In the course of time one would then expect, in the absence of counteracting factors, to get three brachydactylous persons to one normal, but that was not so. There must be other disturbing factors of equal importance.

Punnett was sure Yule was wrong so asked his friend G. H. Hardy about it, thereby drawing Hardy into the development of Mendelian genetics. Hardy, a brilliant mathematician – perhaps the best of his generation – quickly produced a proof by using variables where Yule seems to have been using specific frequencies of Mendelian factors.

The equilibrium principle that Hardy proved was that, after the first generation, allelic frequencies would remain the same for all subsequent generations unless something caused a change, such as selection. Hence in a randomly breeding population of organisms, and in the absence of selection or other causes of change, an equilibrium would be reached after just one generation. For two factors, A and B, the general form of this equilibrium is:

$$p^2AA : 2pqAB : q^2BB$$

We have already explored extensively the factor aspect of this ratio. The three pairs of the factors A and B are AA, AB (or its equivalent BA) and BB. The p and the q are new; p is the proportion of A factors in a population and q is the proportion of B factors. Hardy stipulated that $p + q = 1$. Hence, if $p = 0.4$, q must = 0.6. Let us use those proportions to give Hardy's equilibrium ratio some concrete meaning.

$$p^2AA : 2pqAB : q^2BB$$

becomes

$$(0.4)^2 \, AA : 2(0.4 \times 0.6) \, AB : (0.6)^2 \, BB$$

or

$$0.16 \, AA : 0.48 \, AB : 0.36 \, BB$$

Expressing this in a more familiar way as percentages, 16 per cent of the population will be AA, 48 per cent will be AB and 36 per cent will be BB. Unless there are thing that disturb the equilibrium, the population will have these percentage in every generation. A number of things might cause the equilibrium to be disturbed: natural selection and (yet to be encountered) random drift and meiotic drive, for example. The details and proof of the Hardy equilibrium can be found in Appendix B.

Wilhelm Weinberg independently published similar results in 1908 and articulated the same principle. Consequently, the principle is known as the Hardy-Weinberg principle or the Hardy-Weinberg equilibrium. What is seldom acknowledged is that Karl Pearson stumbled on the same result in 1904. This passage from his 'Mathematical Contributions to the Theory of Evolution. XII. On a Generalised Theory of Alternative Inheritance, with Special Reference to Mendel's Laws' is the relevant one:

Hence, by the above proposition, the distribution of offspring of parents of two couplets is

$$4 \times 4 \times 4 \cdot (\tfrac{1}{4}u + \tfrac{2}{4}v + \tfrac{1}{4}w) \times 4 \times 4 \times 4 \cdot (\tfrac{1}{4}u + \tfrac{2}{4}v + \tfrac{1}{4}w)$$
$$= 4^2 \times 4^2 \times 4^2 \cdot (\tfrac{1}{4}u + \tfrac{2}{4}v + \tfrac{1}{4}w)^2$$

and, by induction, the distribution of offspring for the random mating of parents of n couplets is

$$4^n \times 4^n \times 4^n \cdot (\tfrac{1}{4}u + \tfrac{2}{4}v + \tfrac{1}{4}w)^n.$$

This, except for the constant factor $4^n \times 4^n$, is absolutely identical with the distribution of the parental population, and accordingly if the next generation also mates at random, the mixed race will continue to reproduce itself without change. We therefore reach the following result: –

However many couplets we suppose the character under investigation to depend upon, the offspring of the hybrids – or the segregating generation – if they breed at random inter se, will not segregate further, but continue to reproduce themselves in the same proportions as a stable population.

It is thus clear that the apparent want of stability in a Mendelian population, the continued segregation and ultimate disappearance of the heterozygotes, is solely a result of self-fertilisation; with random cross fertilisation there is no disappearance of any class whatever in the offspring of the hybrids, but each class continues to be reproduced in the same proportions.

Pearson's main purpose in this paper was to demonstrate that Mendelian heredity did not result in new variability that would also be heritable. If, like Yule, he had been aware of the compatibility of Mendelism and biometrics he might have seen the importance of this insight but he did not. Moreover, he failed to notice that the line of reasoning he provides elsewhere in the paper actually demonstrates the enormous variability that recombination creates in a Mendelian system.

The Hardy-Weinberg equilibrium was a crucial discovery. Most scientific theories have an equilibrium principle. For Newtonian physics, it is the First Law: every body tends to remain in uniform rectilinear (straight-line) motion or at rest unless acted upon by an external unbalanced force. Essentially, the law states that if nothing happens then nothing will happen. If a body accelerates or decelerates, one knows that an unbalanced force is at work and the search for it can begin. In Adam Smith's theory of free markets, the metaphor of an 'invisible hand' captures its equilibrium principle. Supply and demand will tend to an equilibrium state and remain there unless factors affecting supply or demand change; if there is a change in supply or demand, things will deviate from the equilibrium. Hence that cause needs to be identified. In mathematical Game Theory, it is the Nash equilibrium, named after its founder John Nash. In a multi-person, repeated game, players will converge on one maximally stable state and remain there unless payoffs in the game change. The game may have more than one equilibrium state, and the state the system ends up in depends on its initial factors.

All these equilibrium principles effectively state that, once equilibrium is reached, if nothing changes, nothing will change. This is important in two respects. First, when a system does undergo change, one knows that some factor in the system has changed. Second, the kinds of factors that one should look for are identified: forces for Newton, factors that increase demand or supply for Smith, factors that have changed payoffs for Nash. The Hardy-Weinberg equilibrium states that the proportion, in a randomly mating population, of Mendel's factors (now known as alleles) will remain constant after the first generation unless factors change the composition of the population (emigration, immigration, selection or mutation, for example) prior to reproduction.

As a side note, although the Danish scientist Wilhelm Ludwig Johannsen introduced the term 'gene' in 1905, it did not come into widespread use until the second decade of the twentieth century.

The adoption of the term 'allele' for what Mendel called 'factors' occurred some time between 1928 and 1931.

Before we pick up other threads of research during the first two decades of the twentieth century, a brief interlude to examine the interrelationship of theory and evidence is in order. As should now be clear, the single important point of division among evolutionists in the 30 years after the publication of *The Origin* was continuous versus discontinuous evolution. Did natural selection act on 'sports' or on small individual variations? Darwin, Weldon and Pearson held that evolution was continuous and natural selection acted on small individual variations, while Galton, Huxley, Punnett and Bateson held that evolution was discontinuous and natural selection acted on 'sports'. Much of the argumentation was mathematical, based on various theories, which involved a significant number of assumptions about how heredity worked. The evidentiary base was not very robust. Galton had performed experiments and collected data on the distribution of traits. Mendel had also performed experiments, the results of which his theory explained exceptionally well, but, as we have seen, his work remained unknown. Prior to 1900 robust evidence about the nature of heredity was scant; most of the work was theoretical and mathematical.

Theories and mathematical derivations are essential in science and we have encountered some central ones, but empirical evidence is equally important. Theory and evidence are not separable. A theory without evidence is disconnected from the empirical world; it might be an impressive edifice of ideas, but there is insufficient reason to believe that the theory actually describes how things in the world work. On the other hand, evidence without a theory is isolated and lacks interpretation. Understanding the inextricable interconnection of theory and evidence is important in understanding the development of evolution, especially in the first two decades of the twentieth century.

Consider a simple, well-known observation. A stick is inserted into a tub of water, so that half is in the water and half is still in the air. What one 'sees' (the actual observation) is a bent stick. Since the stick was not bent when observed before being inserted, some account of what happens at the point of insertion is needed. One could, as a pre-scientific individual might, claim that it was magic. A more rational approach might be to claim that there is some property of water that transforms the stick. On this view (call it an elementary theory of water), the stick is actually bent because an interaction between the stick and the water literally changes the stick. Today, of course, we appeal to a theory of optics to 'correct' the actual observation. The stick is not bent, it just 'appears' that way, and we can explain why using optical theory. It is not the behaviour of water on the stick that is observed, but the behaviour of light when it changes mediums.

Galileo's work on motion provides a richer, more sophisticated example. He rejected Aristotle's physics and cosmology, according to which the stationary earth is at the centre of the universe and the planets revolve around it on spheres, with the outmost sphere containing the stars. Galileo instead championed an alternative theory, the one put forward by Copernicus, according to which the sun is the centre of the solar system and the planets, including the earth, revolve around it. Moreover, the earth moves: it rotates. Galileo provided considerable empirical evidence for this view, including observations of the moons of Jupiter revolving around Jupiter. On Aristotle's theory, the moons would have to crash through the sphere on which Jupiter was located. This was not an irreparable problem. There were ways to modify the theory to accommodate this observation. Taken as a whole, however, the Copernican theory was more mathematically elegant and provided simpler explanations of observations, with few ad hoc assumptions, that is, assumptions made up solely to allow the theory to accord with observations.

Galileo was first and foremost a gifted geometer and he described the movement of bodies in geometric terms. This, in significant part, was why Copernicus's theory appealed to him: it was more mathematically elegant than either Aristotle's theory, which was not mathematical, or that of Ptolemy (90–168 CE), an Egyptian mathematician, geographer and astronomer who was born and died in Alexandria. Ptolemy's theory, like Aristotle's, had the earth as the centre of the universe. Unlike Aristotle's theory, his had the planets revolving around the earth but also revolving around a point on the original circle; sometimes a planet revolved around a point on the second circle. The second revolutions and third revolutions were

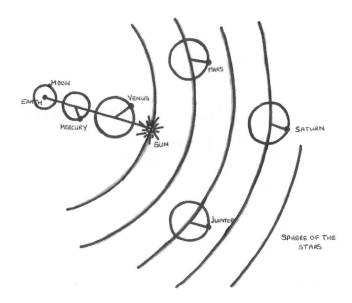

17 Ptolemy's universe.

known as epicycles. His theory was mathematical and explained the motion of the planets much better than Aristotle's theory, but it was very inelegant, with all those circles on circles. Moreover, the centre of the universe was not the centre of the earth. It was slightly off that centre (illus. 17).

Like his predecessors, including Copernicus, Galileo accepted that circular uniform motion was the motion of the heavens. Moreover, he held that circular uniform motion was the perfect motion (unimpeded) of all objects on earth. This was one of the few elements of Aristotle's worldview he retained. His observations, however, did not accord with this. He knew from observations that an object dropped from a tower will fall in a straight line. Furthermore, he observed that it accelerated as it fell. Through a piece of clever geometric manoeuvring, he reconciled these observations with his theoretical commitments. Galileo noted that the object only 'appears' to move in a straight line and to be accelerating. If one reflects on the Copernican theoretical framework he was employing, the earth revolves on its axis. Someone viewing the fall of the object from outside the earth would describe its motion as circular and uniform (non-accelerating). Galileo included this diagram in *Dialogues Concerning Two Chief*

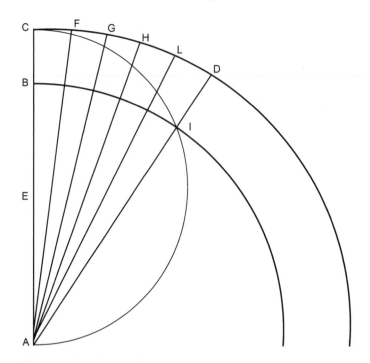

18 Galileo's explanation of the 'actual' path of a ball dropped from a tower. When viewed by someone on earth it 'appears' to fall directly to the ground. When viewed from outside earth, it can be seen to move in a curved fashion because the earth is rotating on its axis. Galileo claimed the curve was a segment of a circle; it is, in fact, a segment of a parabola.

World Systems (illus. 18). This is the case he makes through his character Salviati:

> The first is that if we consider the matter carefully, the body really moves in nothing other than a simple circular motion, just as when it rested on the tower it moved with a simple circular motion. The second . . . it moves not one whit more nor less than if it had continued resting on the tower; for the arcs CF, FG, GH, etc. which it would have passed through staying always on the tower, are precisely equal to the arc of the circumference CI corresponding to the same CF, FG, GH, etc. From this there follows a third marvel – that the true and real motion of the stone is never accelerated at all, but is always equitable and uniform. For all these arcs marked

equally on the circumference CD, and corresponding arcs marked on the circumference CI, are passed over in equal times. So we need not look for any other causes of acceleration or any other motions, for the moving body, whether remaining on the tower or falling, moves always in the same manner; that is circularly, with the same rapidity and same uniformity.

Galileo used the Copernican theory, which he championed. That theory includes a rotating earth. He used that feature of the theory to introduce the concept of the relativity of perspective. This allowed him to explain why, contrary to what we observe, the stone actually moves in a circle (actually, a parabola). Einstein was to take the concept of relativity to its logical conclusion: no perspective is privileged, so all accounts are equally valid. What both examples show is that empirical evidence is interpreted in terms of a theory. This is known as the theory-ladenness of observation and other evidence. We have seen earlier that theories must be consistent with evidence. Here we see that what we 'make of' the evidence – how we understand it – depends on the theories we hold. This is a complex interplay. A theory can be made consistent with the evidence by changing some elements of it or by casting the evidence differently using a new interpretation.

It is time to relate this backdrop to the interpretations of evidence put forward in the first decade of the twentieth century. The understanding of the effect a theory has on how evidence is interpreted is exemplified by the impact that Weismann's germ plasm theory had on the interpretation of experimental results from 1900 to 1910. In 1910 all it took to begin to turn the tide was a simple pointing out that Weismann's theory had no meaningful empirical confirmation. That pulled the rug out from under much of the empirical work that claimed to refute Darwinian gradualism. Indeed, it claimed to refute any role for natural selection.

In the first decade of the twentieth century, the evidence seemed to suggest that natural selection acting on small individual variations was not effective, which gave support to discontinuous evolution. Later, evidence to the contrary tumbled in. Moreover, re-evaluation, using a different theoretical perspective, of the earlier evidence led to its collapse. But this moves us too far ahead in the story. Here, we explore the early work of Hugo de Vries, Wilhelm Johannsen, H. S. Jennings, Raymond Pearl and Elise Hanel. Their experimental work, and the theories of heredity and variation they developed to explain their results, supported discontinuous evolution.

The experimental work of De Vries on selection was considered clear support for his mutation theory. Essentially, De Vries agreed with Darwin that pangenes were the hereditary material but, being influenced by Weismann's experiments and views about germ plasm, he rejected the heritability of environmentally induced changes in organisms. Instead, it is differences in the proportion of pangenes in different individuals that cause the small individual differences seen in populations. Normally the pangenes remained unchanged. Hence natural selection, at best, could affect the proportions but not material itself. This is reminiscent of Jenkin's argument about mutability of species and Darwin's carefully constructed case for the mutability of species in *The Origin*. The difference is that De Vries did believe in evolution and in natural selection; he just did not believe that selection acting on individual differences resulting from stable pangenes was the cause of evolution.

As a result, he held that mutations – alterations in the pangene during cell division – were necessary for evolution. The resulting organisms were sports – macro-mutants. Mutation was, in his view, the necessary source of the variants on which natural selection could act. De Vries's work on hybridization seemed to support his view of the pangene; as noted earlier, it also led to his discovering the same 3:1 proportion that Mendel had discovered when hybrids were crossed. The fundamental difference was that Mendel proposed a theory to explain the ratios, a theory De Vries discovered shortly after his first paper on hybridization, in which he announced the ratios.

De Vries worked with *Oenothera lamarckiana*, a species of flowering plant in the evening primrose family (Onagraceae). Early in his work he had discovered numerous mutations. His subsequent work focused on the stability of a new mutant variety under selection. In his book *Die Mutationstheorie* of 1900, he claimed to have found that seven of his mutant varieties had bred constant and regressed to a new mean. He considered this proof that mutations occur and that selection can promote the mutant, so that a new type regresses to a new mean – evolution has occurred. This evidence, of course, did not refute Darwin's theory that selection acting on small variations was effective in creating new species. It did demonstrate, he asserted, that mutations are an important factor in evolution and, more importantly, supported his theory about heredity and evolution. Consequently, on the basis of his supported theory, he could claim that selection acting on small variations was ineffective because that is what his pangene theory entailed. This provides a clear example of the interconnectedness of theory and evidence.

Wilhelm Johannsen pursued a different experimental path but his work seemed to support De Vries's mutation theory. He focused on what he called 'pure lines', which were descendants of self-fertilization. He reasoned that these would be the most elementary case of heredity transmission. Mendel and De Vries had focused on plants that cross-fertilized and studied hybridization. Johannsen worked with *Phaseolus vulgaris*, the common bean, specifically the Princess cultivar.

One of Johannsen's goals was to investigate Galton's law of regression. That is, over many generations the traits of offspring will move towards the mean of the population. For example, the offspring of very tall parents will slowly, generation by generation, move toward the average height of the population. A crisp summary of this is found in Yule's review of Johannsen's 1903 book describing his experiments and his interpretation of them, 'Galton's Law of Regression [is] the law that the offspring of abnormal parents are *on the average* less abnormal than such parents, "regressing" towards the mean of the race.'

Johannsen was also interested in whether there were types within a population. The approach of Galton, Weldon and Pearson had, for the most part, assumed a homogeneous population, with the regression being to the population mean (average). Johannsen was interested in whether there were independent types within a population. He reasoned that the best method for investigating these features was to develop 'pure lines'. Pure lines, for him, were lineages (generational sequences) where the germ plasm in Weismann's theory or stirps in Galton's theory remain the same in each generation. Self-fertilizing plants that were protected from any possible cross-fertilization gave rise to pure lines. The heredity material would be the same in each generation, whereas cross-fertilization resulted in different combinations of hereditary material in each generation. Pure lines were essential for Johannsen's interpretation of the outcome of his experiment. *Phaseolus vulgaris*, a self-fertilizing plant, was ideally suited for his purposes.

The details of Johannsen's experiments are complicated. It was his interpretation of the results that was embraced in the first decade of the twentieth century and then questioned and abandoned in the second. His experimental method is of less consequence for our story. Johannsen's analysis of his results occurs on two levels. One level is the entire population. What he found was that the results from pure-line breeding conformed to a normal distribution. Moreover, the smallest beans in the F_0 generation yielded small beans and the larger original beans yielded larger beans, but, in both cases, the deviation from the mean was less than in the F_0 generation (illus. 19). Both of these findings supported Galton. The first supported his biometric approach; the second his regression to the mean.

The other level of Johannsen's analysis is of each pure line separately: that is, he examined each line as a separate type within the larger

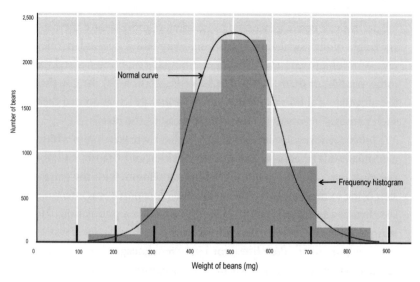

19 Graphic display of the number of beans of each weight.

population. There were nineteen pure lines. His interpretation of the data was that there is at best a weak relationship between the weight of the F_1 bean and the mean and distribution of the F_2 beans. The overall means for the daughter (F_2) generation and the parent (F_1) generation are essentially the same. In addition, the average weight in each group clusters around the overall mean. Hence bean size in the parents (F_1) within a pure line makes no difference to the mean and distribution of the daughters (F_2). Johannsen concluded that the correlation of daughters to parents was zero. A correlation coefficient (a number ranging from −1 to 1) expresses the degree to which two things are associated. The correlation (the word 'coefficient' is usually not added) of lightning and thunder is 1. Lightning never occurs without thunder following and thunder never occurs without being preceded by lightning. Sometimes a person sees lightning but because of the distance the thunder is not heard (a phenomenon sometimes called heat lightning). Nonetheless, there was thunder. Also, sometimes lightning moves from cloud to cloud. It might not be seen, although thunder is heard. Again, however, both did occur together. The correlation of your taking a shower and Great Britain's prime minister meeting with his cabinet is 0. There is no effect of the former on the latter and vice versa. A correlation of −1 describes a complete negative association; the occurrence of one thing ensures the non-occurrence of the other. Since Johannsen was artificially selecting his beans for breeding, a correlation of 0 meant that selection has no effect within type.

In summary, Johannsen's pure line experiments were interpreted as demonstrating four things. First, at the population level, offspring conform to a normal distribution. Second, also at the population level, Galton's law of regression is true. Third, at the individual pure-lines (or types) level, selection has no effect. This last finding supported Huxley's, Galton's and Bateson's view that evolution will not result from selection on small individual variations. It thereby undermined Darwin's view and that of Weldon and Pearson. Fourth, the heritable material is stable from generation to generation. This was consistent with Weismann's germ plasm theory.

Johannsen extended his interpretation to variation. Like Weismann, he distinguished between somatic material and the material of heredity. He held that variation was a somatic phenomenon and that it resulted from the interaction of the hereditary material and the environment. There was no variability inherent in the hereditary material. The soma arose from the hereditary material and was influenced by environmental factors (nutrients available in the soil, which will vary from bean to bean, and position of the pod on the stem, which affects the internal delivery of nutrients, for example). In 1909, after coining the word 'gene' for the hereditary (germinal) material, Johannsen introduced the terms 'genotype' and 'phenotype'. The phenotype is composed of the cells and organs of the organism. The genotype is hereditary material (the genes).

Johannsen continued his pure-line breeding until 1907. This further work, he deemed, confirmed that in pure lines (types) selection had no effect. It established that selecting low-weight beans within a type resulted in the same distribution of weights in the offspring as selecting high-weight beans. Consequently, selection cannot be effective within types. Johannsen's breeding experiments from 1903 to 1907 solidified his view that reversion to the type was complete in every generation. Since low-weight and high-weight beans within a pure line both produce the same distribution in the next generation, there can be no difference in the germinal material of the two beans. A consequence of this is that the germinal material is unalterable and somatic variation is a result of environmental factors acting during the development of the organism.

Johannsen's pure-line method was adopted by Jennings, Pearl and Hanel. They made the modification to the method that their different organisms required. Jennings worked on *Paramecium*, a small (0.5 mm long: smaller than a full stop) single-celled organism. Pearl worked on chickens and studied egg production. Because chickens reproduce sexually, developing pure lines required a slightly different process but the goal was the same. Hanel worked on *Hydra grisea*, a small (at most 10 mm when its tubular length is extended) multi-cellular organism. It reproduces asexually and therefore allows 'pure lines' to be developed and studied. Their results corroborated Johannsen's.

As would be expected, Weldon and Pearson agreed that Johannsen's experiments had added another dimension to be explained. They rejected his interpretations, however, jointly replying to Johannsen's conclusions in an article in *Biometrika* in 1903. Recall that Johannsen claimed that the correlation between the weight of the F_1 bean and the mean and distribution of the F_2 beans was 0. However, he never actually calculated the correlation. He essentially eyeballed the numbers and the absence of a correlation seemed obvious. Pearson and Weldon – one suspects mostly Pearson – actually did the calculation using the data provided and found it to be 0.3481. A correlation of 0.3481, although low, is significantly above zero and far from negligible. The correlation of smoking and lung cancer is in this range. Granted, an overall correlation fails to take into account the fact that the number of cigarettes smoked per day, the length of time one has smoked and concomitant factors such as alcohol consumption all make a difference. Nonetheless, smoking and lung cancer have a correlation of less than 0.5 even when one smokes heavily, has done so for 40 years and drinks heavily.

Yule was also unimpressed. He wrote a review in 1903 in which he summarized Johannsen's finding and conclusions, and, after urging his audience to read Johannsen's book for themselves rather than relying solely on his summary, he also makes the point that the experimental results do not justify Johannsen's claim that selection is ineffective with a pure line. Like Weldon and Pearson, he points out that the effect of selection in the experiment was small but not zero, and if it is not zero then inexorably over time selection can be effective in changing organisms and creating new species. In three passages Yule throws down the gauntlet. The first hints at a potential inconsistency in Johannsen's claim that the germinal material remains unchanged and yet mutations occur:

> In the second place one must consider 'Mutations, or the possibility of discontinuous variations of the type.' 'That they do occur' says Professor Johannsen, 'appears to me beyond doubt.' The latter statement seems to indicate that the writer does not consider the existence of mutations would invalidate his theory.

In the second, he focuses on burden of proof. In his view, given the evidence and his interpretation of it, the burden of proof rests with those who want to interpret the data as demonstrating that the effect of selection is zero. Note again the interplay between evidence and interpretation. Johannsen's theoretical commitments are different from Yule's, which results in a difference in what they make of the data.

> It seems difficult then, on very wide grounds, to admit that the effect of selection within a 'pure line' or the intensity of heredity within such a line can be rigidly zero, *i.e.* the 'burden of proof' lies with those who hold such a conception, which is inconsistent with the conception of evolution itself.

In the third, he returns to mutations: 'If Professor Johannsen believes in the occurrence of mutations he ought to believe in the effect of continued selection within the race, whether accepting the hypothesis of continuous variation or not.'

Weldon and Pearson were committed biometricians who also accepted continuous evolution; Yule was somewhat more neutral and consequently should have been more persuasive. All the criticisms of the interpretations of pure line research, however, were lost in the dazzle of the methodological brilliance of Johannsen's experimental work, and of those who followed his method. Most biologists embraced that method, the theoretical assumptions he made and his conclusions. By 1910, with the additional work of Johannsen himself, along with that of Jennings, Pearl and Hanel, almost all biologists had been convinced that evolution was discontinuous and selection was not a major mechanism of evolution. Jennings interpreted his work on *Paramecium*, for example, as supporting Johannsen's theory of heredity. Moreover, it supported Johannsen's claim that variations in the soma (the phenotype) were due to the environment. Hanel's on *Hydra grisea* yielded the same results as Johannsen's and was widely regarded as adding great strength to his views. Pearl's work on chickens showed no effect of selection on egg production over a nine-year period.

During this period, Pearson continued to identify flaws in the research and the reasoning. He was mostly scorned and/or ignored by biologists. Pearson was a mathematician; few biologists had any sophistication in mathematics and most focused narrowly on biological experiments. It was a dark moment for mathematical theorists and for Darwinians. But the dawn of the second decade of the twentieth century brought a ray of light for both. In late 1910 the tide began to turn, in part because the flaws in Johannsen's experimental designs were exposed. Arguably more important, however, was the fact that the assumptions made on the basis of theoretical commitments were shown to be untenable. The mathematical reasoning and underpinning of their positions were demonstrated to be flawed. This opened up the space for a new line of experimental research to flourish. That research reversed nearly all of Johannsen's interpretation and impugned the experimental methods on which they were based.

The watershed point in 1910 was, in the view of many historians, J. Arthur Harris's address at the American Society of Naturalists at the end of the year. His address was part of a symposium on pure lines and genotypes. His assessment of the empirical work using pure lines was neutral, and he neither agreed nor disagreed with the experimental work. His focus was on the presuppositions made by the pure line experimenters. Harris's key point, one that went to the heart of the problem with De Vries's mutation theory and the pure-line research, was that a stable, unchanging germ plasm was assumed. De Vries and the pure line researchers had adopted almost entirely Weismann's distinction between germ plasm and soma, and his claims that the germ cells remain constant during development from a single cell to the adult organism and that untransformed germ cells were transmitted to the next generation. These claims were postulated on the basis of his observation of a lineage of germ cells of hydrozoans. In that organism's development from a single cell (the fertilized ovum) to the mature adult, the germ cells remained constant. There are, as Harris pointed out, many other potential explanations of this observation. Weismann's, however, became the currency of biological research until 1911. Theoretical commitments, Harris had noticed, resulted in the privileging of one of many possible interpretations of the empirical data.

Some work done in the later years of the first decade was already unravelling the evidentiary and, most significantly, the theoretical basis of the pure-line research. The postulates of Weismann's theory began to teeter. In 1911 that theoretical framework began a complete collapse. Further work after 1910 simply confirmed its inadequacies and the need for a different theory.

William Ernest Castle (1862–1962), C. B. Davenport (1866–1944), T. H. Morgan (1866–1945), H. Nilsson-Ehle (1873–1949), Edward Murray East (1879–1938) and H. J. Muller (1890–1967) were all crucial players in the conversion of biologists and theorists. By the end of the 1920s, conversion to continuous evolution was virtually universal. As an example, by 1916 Jennings had converted. Although Pearl never converted, he was in the minority by 1918.

Castle began as an advocate of the pure-line view and discontinuous evolution. He had no formal background in mathematics. His doctoral work was in embryology. Nonetheless, his early views were influenced by reading Bateson's *Mendel's Principles of Heredity: A Defence*. He was not competent enough mathematically to assess the consequences of Mendel's theory. He was also not competent to assess the biometric literature. As has been pointed out, Bateson was not competent either. Castle's adoption of Mendelism was an adoption of Bateson's Mendelism. This amounted to one drunk holding up another. Castle and Pearson clashed in 1903 when Castle published 'The Laws of Heredity of Galton and Mendel, and Some Laws Governing Race Improvement by Selection'. The paper was hopelessly confused on correlation and regression, and Pearson dismantled his arguments and threw down the gauntlet. Essentially he demanded that Castle provide a mathematical proof of his claims. Castle never did.

Castle worked with rats. The trait in which he was interested was coat colour in hooded and Irish rats (illus. 20). The pattern and colour vary.

In 1906 Hansford MacCurdy, working in Castle's laboratory, had completed experiments studying coat colour and distribution in these rats when they were cross-bred. The introductory comments in their publication of these results in 1907 ('Selection and Cross-breeding in Relation to the Inheritance of Coat-pigments and Coat-patterns in Rats and Guinea-pigs') make it clear that Castle had changed his view. He now saw Mendelism and Darwinian selection as compatible.

Bateson [in a 1903 article] refers to Crampe's failure to obtain the self or the albino conditions by selection from the particolored, and adds: 'The types are in fact definite, and can not

20 A brown hooded rat.

be built up by cumulative selection.' The statement applies strictly only to the two extremes of the series, viz, self and albino, but by implication to the others also. Our experiments, however, indicate that it is possible to modify by selection and cross-breeding both the Irish and the hooded conditions, leading to the production of intermediate conditions. We suspect that the same may be true of the self condition also . . .

The theoretical importance of this is obvious. Cross-breeding and selection combined are means by which we may not only modify existing Mendelian characters, but may even create new ones. They are, then, factors of prime importance in evolution, even in the case of characters which vary discontinuously.

They also make clear later in the paper that their results were inconsistent with De Vries's views:

De Vries and the Darwinians differ not only as to the part which selection plays in evolution, but also as to the nature of the material upon which selection acts. According to

De Vries, species are not modified by selection; mutations are new species and selection determines only what mutations shall survive, fluctuations having no evolutionary significance. On the Darwinian view, all species, whether arising by mutation or not, are subject to modification by selection . . . So far, then, as these experiments go, they support the Darwinian view rather than that of De Vries.

Furthermore, MacCurdy and Castle undermine the biometricians' law of regression:

One of the noteworthy features of the case is the absence of what may properly be called regression. The filial average does not, with any uniformity, lag behind the parental average, in the process of displacement downward in the variation figure. This fact, together with the decreasing skewness of the variation curve, indicates that the effects of selection in reducing the extent of the pigmentation will be permanent, that is, that a stable, narrowstriped variety of hooded rats can be established by selection, and that this variety will breed true.

These were important experimental results. The experimental tide was turning toward Darwin. First, Mendel's theory could explain the hereditary nature underlying the results, meaning that Darwinism and Mendelism were compatible. Second, the results cast doubt on the biometricians' favoured law of regression. Whatever value the biometrical approach may have had, the law of regression was not part of it.

The year after MacCurdy and Castle published their results and interpretation of them, Nilsson-Ehle published the preliminary account of his results in crossing oats. He investigated the colour difference in glumes – a husk around the oat seed. He discovered that the ratio of black glumes to white was 15.8:1 in the F_2 generation. He was familiar with Mendel's experiments and theory and interpreted his results using it. His 15.8:1 ratio is very close to what one would predict based on Mendel's theory. Using Mendel's theory,

Nilsson-Ehle calculated the expected ratio for two-factor segregation, which is 15:1. He continued his crossing experiment with wheat. In 1908 he published a brief account and in 1909 a reasonably comprehensive account of his work, his findings and the consistency of his findings with Mendel's theory.

The importance of his work was immediately obvious. It confirmed what Mendel had already suggested about traits that were controlled by multiple factors. But Nilsson-Ehle's results justified extending the Mendelian system to many more factors (ten separate factors would give rise to more than 50,000 forms) and that made plausible Nilsson-Ehle's conclusion that 'many mutations, above all in exotic plants, are only new groupings of already present factors and really represent nothing new, especially in such cases where they throwback'.

Mendel's theory allows the number of variations in a multi-locus two-allele-per-locus system to be generalized. The number of variations equals 2^n when dominance is complete and 3^n when dominance is incomplete, where n is the number of loci. Incomplete dominance is an addition to Mendel's original system and evidence to date suggests it is pervasive. Complete dominance occurs when one allele dominates over the other such that the heterozygote has exactly the same trait as one of the homozygotes – namely, the homozygote with the two dominant alleles. Incomplete dominance occurs when the heterozygote can be distinguished from both homozygotes. There are thus three different and distinguishable traits. Hence a 30-locus system with incomplete dominance yields 205,891,132,094,649 (about 206 trillion) varieties. Clearly there is more than enough variation in a population from segregation and recombination alone to provide a basis for selection. There is no need for mutation. The mystery of the maintenance of variation is solved. Continuous evolution now looks a lot more plausible.

Edward Murray East, working on maize in the U.S., initially considered that the results of his early experiments, which showed considerable variation, were consistent with the results and interpretation of Johannsen, Jennings and Pearl, but later sided with Nilsson-Ehle on the incredible amount of variation from segregation and recombination and the effectiveness of selection on this variation.

Like many others, Thomas Hunt Morgan, an influential and brilliant geneticist, was also initially dismissive of natural selection as the key mechanism of evolution. Interestingly, he was also dismissive of Mendel's theory. Morgan worked with fruit flies (*Drosophila*). That work made him increasingly sceptical of prevailing theories and Bateson's interpretation of Mendel's theory. Contrary to what he initially thought, his experimental results could be more naturally interpreted in terms of Mendel's theory. He slowly came to see this. As a result, by 1910 he too had become a convert, accepting Mendel's theory and the efficacy of natural selection. One key outcome of his genetic experiments was evidence that very small variations behaved in accordance with Mendel's theory and hence could not be swamped within a population. Jenkin's swamping theory had been laid to rest.

The tide had turned and Darwinian gradualism had become the dominant view by the end of the second decade of the twentieth century. This experimental part of the story plays a key role in understanding the rejection and then restoration of Darwinian gradualism as well as the emerging acceptance of the consistency of Darwinian gradualism and Mendel's theory. What clinched the entire matter was the mathematical integration of Darwin's theory with Mendel's. As a bonus, the mathematical work that brought about that integration in the 1920s also integrated biometrics.

SIX

A Bridge over Troubled Water

Ronald Aylmer Fisher (1890–1962) was a leading mathematician of the first half of the twentieth century. Along with J.B.S. Haldane (1892–1964) and Sewall Wright, he shaped modern population genetics. But there are many other legacies of his pioneering work. The one most widely encountered involves experimental design. Fisher championed randomized controlled trial (RCT) in his book *The Design of Experiments* (1935). There he set out the RCT method, and argued that randomization, control and replication reveal causal relations.

Today his method is pervasive in clinical medical research, but Fisher primarily used this method in his agricultural experiments. In that domain, for example, the method can be used to determine the effect of nitrogen fertilizer on maize growth, health, yield and the like. A field is divided into blocks. Adjacent blocks are paired: see the chart overleaf, in which each A-block is paired with the B-block directly below it: the same for C-blocks and D-blocks. One block from each pair is randomly assigned to receive the fertilizer (the intervention); the other block is the control (no fertilizer) (illus. 21). Selecting randomly which block receives the fertilizer removes bias; it also makes statistical analysis of the results possible, but that is a complicated matter. At the end of the season the blocks can be compared. If the blocks receiving fertilizer grew faster or were healthier or had higher yields than the control, then, Fisher claimed, one could declare with confidence that the fertilizer caused that outcome.

This method is ideally suited to agricultural experiments because the plants are bred to be homogeneous in terms of important traits. It is also highly probable that adjacent blocks in the field will be

Blocks A	C	F	F	C
Blocks B	F	C	C	F
Blocks C	F	F	F	C
Blocks D	C	C	C	F

21 A-blocks are paired with B-blocks (as illustrated in the grey-shaded blocks) and C-blocks are paired with D-blocks. One block of the pair is chosen at random to receive fertilizer; the other is the control, as in the line-shaded blocks.

homogeneous for all the factors that affect plant growth: amount of rainfall and sunlight, and humus levels and texture, for example. In clinical medical research, however, the assumptions of homogeneity do not apply; humans are heterogeneous for all traits. Moreover, randomization – which, it is claimed, tames the heterogeneity – is messy in clinical settings and seldom achieved in the mathematically required way. Hence the extension of his method to clinical medical research is suspect. Nonetheless, it is a brilliant design for the context in which he principally used it.

Fisher's contribution to evolutionary theory is of greater interest here. Through his work and that of Haldane and Wright population genetics was born. As a result, evolutionary theory was mathematized in a way that integrated Mendel's theory, Darwin's theory, with its gradualist assumption, and biometrics. The focus on Fisher in this chapter results from the fact that he provided a comprehensive account of population genetics and evolution in a single volume, making it easier to see the assumptions, his arguments and his conclusions. (The contributions of Haldane and Wright were very important as well. In the next chapter, the importance of Wright's contribution will be highlighted.)

Fisher made clear his goal at the outset through the choice of title of his book of 1930, *The Genetical Theory of Natural Selection*. This signals that his goal is to show that genetics (Mendelian genetics) and Darwinian natural selection (gradualism) are compatible and that a unified theory embracing both can be constructed, which consequently demonstrates the compatibility.

Fisher also signals in one of the epigraphs to chapter Two that he intends to provide a rigorous unified theory, like those in the physical sciences. The epigraph is from T. H. Huxley's article of 1854, 'On the Educational Value of the Natural History Sciences': 'In the first place it is said – and I take this point first, because imputation is too frequently admitted by Physiologists themselves – that Biology differs from the Physico-chemical and Mathematical sciences in being "inexact".' Huxley vigorously denies that biology is a less exact science in either methods or results than the physico-chemical sciences. There is a difference between the mathematical sciences and both the physico-chemical and biological sciences; but even that is a difference of degree rather than kind. A little later in this article, Huxley wrote (emphasis added):

> The Mathematician deals with two properties of objects only, number and extension, and all the inductions he wants have been formed and finished ages ago. He is occupied now with nothing but deduction and verification.
>
> The Biologist deals with a vast number of properties, and his inductions will not be completed, I fear, for ages to come; *but when they are, his science will be as deductive and as exact as Mathematics themselves.*

By providing an epigraph from Huxley's article, Fisher connects his task with the sentiments of Huxley. He, Fisher, is going to render an exact scientific account of evolution. That this is his goal is reinforced by his numerous comparisons of the biological formulas he constructs with economics (specifically principal, interest and repayments of loans) and with thermodynamics. Moreover, statements like 'A similar convention, appropriate in the sense of bringing *the formal symbolism of the mathematics into harmony with the biological facts,* may be used with respect to the period of gestation' make explicit his goal of mathematizing evolutionary biology and thereby unifying Mendel's theory with Darwin's.

His unification is achieved, for the most part, in the first two chapters. In later chapters he provides important amplifications of some key elements but the essence of his theory is contained in chapters One and Two. His formalization, it is worth noting, does not depend directly on the empirical work examined in the previous chapter.

That work cleared away some impediments but Fisher is pursuing the formal characterization of a theory that will explain empirical evidence. He aims to give a theory of evolution that mirrors Newton's physical theory.

By now it will be obvious that formalizing a theory requires providing a formal statement of its axioms. Darwin never did that in *The Origin*; his was an informal statement of what he took to be the fundamental laws. Fisher had to mathematically formalize the relevant axioms of Darwin's theory, although he does not specifically refer to Darwin's fundamental laws. Nonetheless, the effect of his arguments and formal (mathematical) account in chapter Two does formalize them. In addition, Fisher's unification of Mendel and Darwin required an additional axiom.

There have been many debates about, and refinements to, Fisher's formalization. In addition, there have been many misinterpretations. Many of these, as Jean Gayon has noted, arise from Fisher's own somewhat idiosyncratic use of expressions such as 'fitness of any organism'. Nonetheless, two things are beyond dispute. First, the internal structure of his theory is clear. Second, the explanatory and predictive consequences of the theory are determinable; that is, they are mathematically deducible from the theory.

Fisher begins by clearing away two rotten conceptual beams that still lingered in discussions of evolution. Blending heredity, which the Mendelians correctly rejected, was one rotten conceptual beam. The evolutionary importance and efficacy of mutations, which the biometricians Weldon and Pearson correctly rejected, was another. Both are dispatched deftly in chapter One. In place of blending heredity he assumes particulate heredity. That assumption is the foundation of Mendel's theory. In place of mutation as fundamentally important he demonstrates that selection acting on small individual variations is an effective mechanism of evolutionary change. That position is the foundation of Darwin's theory. Thus the stage is set for the unification of Mendel and Darwin.

Having removed these rotten conceptual beams, he turns to the task of constructing the theory. He identified three things that required attention. They are old friends of ours: the struggle for existence (resulting in natural selection), heredity and the nature of variation.

Heredity was readily provided by Mendel's theory, which Fisher adopted completely. The main task before him was to integrate natural selection and variation, which he deduced from the struggle for survival, into Mendel's theory. Accomplishing this meant that his task was done.

He began with some axioms of Darwin's theory:

Axiom 1: The demarcation of species is artificial (organisms manifest an insensible gradation of forms).

Axiom 2: There is a struggle for existence among organisms (the conditions of existence).

Axiom 3: There exist small trait-differences among organisms (variation).

Axiom 4: Small trait-differences are heritable (heredity).

Axiom 1 is a *sine qua non* (that without which nothing). Without it evolution cannot occur. Darwin provided a compelling nominalist-style argument for its acceptance. Most evolutionists on all sides of the debates, thereafter, took this for granted. Fisher also simply assumed it. Since axiom 4 is already expressed mathematically and proved by Mendel's theory, Fisher concentrated on a mathematical specification of axioms 2 and 3.

Fisher used 'the life table' as the starting point for exploring axiom 2. The life table had long been used by insurance companies and demographers. Fisher simply expropriated it: 'In order to obtain a distinct idea of the application of Natural Selection to all stages of the life-history of an organism, use may be made of the ideas developed in the actuarial study of human mortality.'

Although these are euphemistically called life tables, they are really death tables, just as life insurance is actually death insurance. In addition to death rates, Fisher used reproduction rates in express-ing mathematically axiom 2. His goal was:

The object of the present chapter [chapter Two] is to com-bine certain ideas derivable from a consideration of the rates of death and reproduction of a population of organisms, with the concepts of the factorial scheme of inheritance, so as to state the principle of Natural Selection in the form of a rigorous mathematical theorem, by which the rate of improvement of any species of organisms in relation to its environment is determined by its present condition.

This, in effect, is the construction of a mathematical expression of population expansion (or decline), which leads to a struggle for survival. The death and reproduction rates of organisms are first given by separate equations and then combined into a single equation, which is his axiom 2 (the technical details can be found in Appendix B). For Fisher, as for Darwin, natural selection was a deductive consequence of the struggle for survival. Hence natural selection has been proved.

Fisher now turned to axiom 3. Mathematically specifying this axiom requires axiom 4. Mendel had already provided a mathematical expression of this axiom and empirically proved it. Fisher explored its implications with respect to variation. He first compared blending inheritance and Mendelian particulate inheritance with respect to variation: 'It has not been so clearly recognized that particulate inheritance differs from the blending theory in an even more important fact. *There is no inherent tendency for the variability to diminish*' (emphasis added). This, of course, was what Bateson and many others had failed to realize. Fisher then expressed what we know as the Hardy-Weinberg equilibrium (strangely, he made no reference to Hardy or Weinberg even though their publications had appeared 22 years earlier):

In a population breeding at random in which two alternate genes of any factor, exist in the ratio p to q, the three genotypes will occur in the ratio $p^2 : 2pq : q^2$, and thus ensure that their characteristics will be represented in fixed proportions of the population, however they may be combined with characteristics determined by other factors, provided that the ratio $p : q$ remains unchanged.

If the ratio remains unchanged, variation remains undiminished. As is frequently the case in his writings, Fisher assumed that the equilibrium, its proof and its implications for variation would be obvious to anyone with a modicum of mathematical ability, so he provided no proof. This equilibrium is an obvious mathematical consequence of the particulate theory of heredity (Mendel's theory). Fisher was clear that chance survival and natural selection were among the factors that would change the ratio $p : q$.

This equilibrium constitutes another axiom. It needs to be added to Darwin's axioms, and is crucial to a unification of Darwin's theory and Mendel's theory. This axiom can be stated succinctly:

Axiom 5: If the ratio of two alleles a_1 and a_2 at a locus is $p : q$ in F_0 generation, the alleles will be distributed $p^2(a_1) : 2pq(a_1a_2)$: $q^2(a_2)$ in all subsequent generations, F_I–F_∞, unless something disturbs the ratio.

The struggle for existence, which results in natural selection, variation and Mendel's particulate theory are now all formulated mathematically.

What remained unresolved was how variation was maintained when the equilibrium state was disturbed by natural selection or some other factors. To resolve this, he brought biometry into the fold. He provided a mathematical (statistical) treatment of ratios and the effect of gene substitutions for quantitative (biometric) traits such as height. Although remarkably insightful, the account is compact and not easy to follow. In his enlarged version of this book published in 1958, he provided a more detailed mathematical account. Height was prominent in the biometric analyses of Galton and his followers, so it was not an accident that Fisher explored this specific trait. This connects biometry to the theories of Mendel and Darwin.

With these axioms in place Fisher explored further natural selection, focusing on fitness. Natural selection was one important mechanism that disrupted the equilibrium of the Hardy-Weinberg equilibrium. Fisher set up his formalization simply (the technical detail can be found in Appendix C). Fitter organisms, Fisher noted, on average, would survive. Less fit organisms would not survive.

This needed to be incorporated into the heredity formalization, which Fisher did (the technical details are in Appendix D).

This led to his Fundamental Theorem of Natural Selection: *The rate of increase in fitness of any organism at any time is equal to its genetic variance in fitness at that time.* This he compared to the second law of thermodynamics and maintained that it, like that law in physics, held 'the supreme position among the biological sciences'.

His presentation of the theorem is very cryptic. As a result, it has been the subject of much controversy. Many have assumed that he was claiming that population fitness would always increase, which is almost always going to be false. It is only true of single-locus selection. James Crow has provided a more nuanced and quite defensible interpretation of the Fundamental Theorem. He claims that the theorem is restricted to additive genetic variance, with no claim about population fitness. However it is interpreted, Fisher's theory can survive without it.

What is important about Fisher's theory is that it resolved several of the controversies that occurred in the 60 years after the publication of *The Origin*. The central one was over continuous versus discontinuous (saltational) evolution. Fisher demonstrated that evolution is continuous with natural selection acting on small individual variations. Just as Darwin insisted to his death. Darwin's informal exposition of the theory was not up to the task of settling the controversy. Fisher's was.

Haldane and Wright also made important contributions to evolutionary theory. They concentrated on different aspects of population genetics and evolution. They, like Fisher, accepted Darwin's view that natural selection acted on small individual variations. Haldane found clear evidence that alleles at loci close together on a chromosome are inherited together. They are linked. The explanation draws on another feature of chromosomes: crossing over, a phenomenon that had been studied extensively by T. H. Morgan in *Drosophila* (fruit flies) during the second decade of the twentieth century. Crossing over occurs when two chromosomes exchange segments. When a chromosome remains 'intact' – no crossing over – all the alleles on that chromosome will be inherited together; to reiterate, they are linked. When, however, two chromosomes exchange segments, the alleles on the exchanged segments are no longer linked to the alleles on the original chromosomes. Haldane's theory allowed him to determine, mathematically, the numerical values (the percentages) for crossing over in various circumstances (illus. 22).

Using these values, he was able to explain why alleles close together on a chromosome have a higher linkage value. In essence, he was able to deduce from his theory that as the distance separating alleles increases, the probability that they will become separated via crossing over also increases. Although this seemed intuitively obvious to biologists, Haldane was able to prove it mathematically.

Wright's contributions were for a long time underappreciated. His contributions included developing an analytical method (which he called path coefficients), introducing the concept of adaptive landscapes and introducing random drift as a cause of some evolutionary changes (illus. 23). The method of path coefficients allowed him to examine genetic causes of traits and determine the effect of various causes on a trait. Causes can be separated and potentially quantified. The use of adaptive landscapes is an analytic tool for visually displaying the relationship between genotypes (the genetic makeup of an organism) and success in reproduction. This tool can also be used to explore the relationship of phenotypes (the physical properties of a whole organism) and success in reproduction. The higher the peak

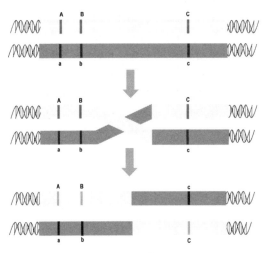

22 Crossing over of chromosomes.

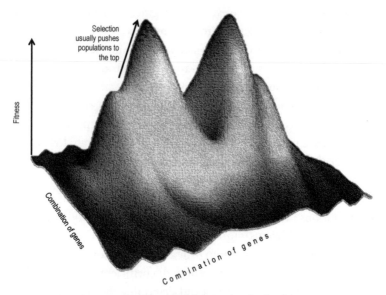

23 An example of Sewall Wright's adaptive landscapes.

in a landscape, the greater the fitness (reproductive success) of the organism. In the next chapter, we will examine random drift.

In the hands of Fisher, Haldane and Wright, evolutionary theory was mathematically formalized and Darwin's theory, Mendel's theory and biometry were combined into a single theory. That single theory had been shown by the 1930s to be able to provide powerful explanations of observed phenomena, and to provide robust predictions, which guided future research. This was the first evolutionary synthesis; that is, it was the first bringing together – synthesizing – of separate fields. The next synthesis brought almost all the branches of biology together under the theoretical framework of evolution.

SEVEN

Evolutionary Biology
Comes of Age

By 1935 the task of providing an *initial* mathematical formalization of evolutionary theory was completed. Evolutionary biology
now had a robust formal theory. There was still much work to be
done. Working out the implications of the theory for understanding
biological phenomena was essential. The theory explained and
predicted changes in gene frequencies in populations, especially
changes due to natural selection. It is natural selection that makes the
theory a theory of evolutionary change.

The relevance of the theory to explaining the incredible diversity
of species seemed clear. The formation of new species (speciation)
involves natural selection acting on genetic variation and that is
exactly what evolutionary theory is about. The fact that the theory
was relevant to explaining the biogeographic distribution of organisms also seemed clear. Exactly how speciation works and how that
explains biogeographic distribution needs some elaboration, however.
That comes later in our story.

The theory's relevance to other areas of biology was less evident.
The relevance of evolutionary theory to embryology, for example,
was not obvious. It seemed unlikely that the theory could shed light
on the development of a fertilized ovum into an adult. It appeared
more likely that discoveries in embryology might shed light on some
of the factors that led to changes in gene frequencies, although few
examples were available. The same asymmetry appeared to be true for
palaeontology and taxonomy. Palaeontologists at this time were
inclined to interpret their evidence as more consistent with discontinuous evolution. As a result, palaeontologists tended to reject both

Darwin's version of natural selection and Mendel's theory, which was almost universally regarded as supporting Darwin's view.

On the surface, the relevance of evolutionary theory to genetics should be straightforward. After all, Mendelian genetics was an essential pillar of the theory. In 1935, however, it was seen as only partially relevant. Some explanations and predictions in genetics were based on Mendelian genetics (genetic diversity in populations, for example), but a large amount of genetic research focused on identifying genes, or complexes of genes, in specific organisms, and on mapping genes to traits. The utility of Mendelian genetics to that endeavour was not obvious. For that reason, much of biology was disconnected from evolutionary theory in 1935.

In 1962 Thomas Kuhn published his now-famous *The Structure of Scientific Revolutions*. In it he introduced the idea of a paradigm in science: a framework into which individuals in a domain of science are inculcated and become committed. A paradigm is more than the currently accepted theory, but its central element is an accepted theory. Importantly, it provides the social glue that gives a domain its professional cohesion. Although evolutionary theory had impressively unified three research domains – Mendelian genetics, Darwinian selection and biometrics – it had not yet become a biological paradigm.

This, in part, was due to another feature of biology in 1935. The various domains of biology lacked a common methodology. In addition, plants (the domain of botany) included a vast array of non-sexually reproducing species, which Mendelian heredity did not encompass. Mendel's research was on sexually reproducing plants. As a result, evolutionary theory in 1935 was relevant only to sexually reproducing plants and animals. Consequently biology was far from a 'profession', since it lacked the required cohesion. There were obvious interconnections between many domains of biology: the connection of taxonomy to palaeontology, for example. It is difficult to imagine research in palaeontology that did not assume some system of classification. Equally difficult is imagining comprehensive research on classification (species, genera, families and so on) without paleontological data. The connection between zoology and botany is also obvious. Many plants, like almost all animals, reproduce sexually and the changes in their gene frequencies are explained by population genetics. So why would the study of sexually reproducing plants be disconnected from that of sexually reproducing animals?

Although there seemed to be numerous interconnections between the different domains of biology, there was no unifying theory and no common methodology. Creating a unified biology, and in the process creating a 'profession of biology', required the concerted effort of key figures in almost all the sub-fields of biology during the next twenty or so years. Creating a single cohesive discipline of biology required a unifying theory. Evolutionary theory became that unifying theory. During the next 25 years, the domains were united with evolutionary theory one by one, a process that resulted in a deepening and strengthening of evolutionary theory. The population genetical theory of Fisher, Haldane and Wright was the starting point. By 1960, however, it had been transformed and had become one among many components of the contemporary theory of evolution. The process of unifying biology was largely complete by 1960. By 1973 'Nothing in Biology Makes Sense Except in the Light of Evolution', the title of an article in *American Biology Teacher* by a participant in the unification, the geneticist Theodosius Dobzhansky, was a truism.

The culmination of this unification and professionalization was dubbed by another of those involved in bringing it about, Julian Huxley, as 'the modern synthesis'. The resulting theoretical framework became known as the modern synthetic theory of evolution. It completed what Darwin sketched and it provided a comprehensive mechanistic account of evolution. Moreover, it brought almost all the domains of biology into the evolutionary tent. This completed Darwin's quest for Whewell's consilience of inductions. It demonstrated that evolutionary theory was explanatory in, and central to, all domains of biological inquiry. But we are moving ahead of the major events that wrought this unification. These events are integral to the evolutionary story. The political and scientific dynamics at play in the period from roughly 1935 to 1960 raise the curtain on an exciting drama.

As in all dramas, the players are a key component. Unlike dramas on the limited stage of a theatre, however, the players in this drama are on the stage of life, which includes the far-flung corners of the globe. Some sections of the stage (the United States, Britain and Germany) loom large in the drama, as do a few players: the geneticists Theodosius Dobzhansky and E. B. Ford (1901–1988), the palaeontologist G. G. Simpson, the taxonomist (systemicist) Ernst Mayr, Julian Huxley (1887–1975), the cytologist Michael J. D. White, the botanist G. Ledyard Stebbins and Bernhard Rensch (1900–1990).

Dobzhansky was one of the twentieth century's most influential geneticists and the first to advance the unification. He was born in 1900 in what is now Ukraine (then part of the Russian empire). He decided early in life that he wanted to be a biologist and in 1917 entered the University of Kiev and received his degree in 1921. At that time, he was working on beetles (specifically the family Coccinellidae, commonly known as ladybirds or ladybugs). He stayed at the university until 1924 as an instructor in zoology. At some point, probably in 1923, he obtained some fruit flies (*Drosophila melanogaster*) while in Moscow. H. J. Muller had imported the stock to the Soviet Union. He turned his attention to *Drosophila* research, discovering pleiotropic effects (genes that produce more than one phenotypic trait). He published his findings in an article, 'About the Construction of the Sexual Apparatus of Some Mutants of *Drosophila melanogaster*'. In 1924 he took up a lectureship at the University of Leningrad and continued to work on *Drosophila melanogaster* and pleiotropy. In 1927 he received a fellowship from the Rockefeller Foundation and moved to the United States, where he worked with Thomas Hunt Morgan at Columbia University. Three years later he moved with Hunt to the California Institute of Technology (CalTech), where he remained until 1940, when he returned to Columbia. Some of his later work was on the fruit fly *Drosophila pseudoobscura*. Numerous leading geneticists studied under him, including Richard Lewontin and Francisco Ayala. In 1975 he died of heart failure after a seven-year struggle with lymphocytic leukaemia.

Dobzhansky's major contribution to the synthesis was his book *Genetics and the Origin of Species* (1937). Fisher, Haldane and Wright had provided the mathematical formalization for the initial synthesis. What was needed now was a demonstration that empirical phenomena in the diverse domains of biology could actually be explained or predicted by the theory. Science without theory is fragmented and knowledge collection is akin to stamp collecting, to borrow a powerful image from Dobzhansky. Theory without explanatory and predictive connection to phenomena is a quaint, if interesting, mathematical system devoid of empirical meaning. Deducing from the theory the behaviour of empirical systems, which correspond to observational and experimental results, is essential. *Genetics and the Origin of Species* provided the empirical touchstone of the validity of population genetics.

The connection of the behaviour of chromosomes and Mendel's theory had been made early in the twentieth century. Dobzhansky's research solidified that connection. His research on the genetics of fruit flies revealed the connections between genes and traits. It also demonstrated that genetic variability within a species was vastly greater than had been assumed. Four new findings emerged from his research. First, mutations occur frequently. Some are deleterious and the organism will not be viable, but many are neutral and will be preserved in future generations. Since mutations will vary within a breeding population, there will be enormous variation. Second, populations of the same species that are reproductively isolated (do not interbreed) will have a different mutational variation. Third, some neutral mutations will become advantageous in changed circumstances (changes in the environment) and will spread throughout the population over generations. Fourth, populations that remain reproductively isolated for long periods of time will have different mutations and eventually will become reproductively isolated. Two individuals, one from each population, will not be able to produce viable, non-sterile offspring. Dobzhansky collected flies from a wide geographic area of North America and demonstrated the empirical validity of each of these points. Some demonstrations came from direct observation of the chromosomes of the collected flies. Others, such as the spread through the population of genes that were originally neutral and then given a selective advantage, were demonstrated experimentally.

These results were powerfully employed in *Genetics and the Origin of Species* to substantiate two foundational aspects of Darwinian evolution. The first is that small changes in the genes of individuals occur frequently and ensure the constant generation of variation, and that this incredible reservoir of variation allows organisms to adapt to changed environments. Neutral mutations became selectively advantageous in changed environments. Second, within a species, two or more populations of organisms that are reproductively isolated will accumulate different mutations from each other, and will eventually be reproductively incompatible, at which point they will be two species. This provides the empirical support for the theoretical account of the origin of species. In one sweep, Dobzhansky showed that understanding the genetics of organisms requires evolutionary theory and that evolutionary theory requires the evidence provided by genetic research, and they are entirely compatible. Genetics was brought fully into the evolutionary fold.

Ernst Mayr was born in 1904 in Bavaria. He received his PhD in zoology from the University of Berlin in 1925 and subsequently, with funding from Lord Walter Rothschild, he travelled to New Guinea. Initially he was located in Dutch New Guinea (the western part of the island and part of the Dutch East Indies). When his collecting commitment ended, he travelled to the Mandated Territory (the northern portion of the eastern part of the island), which had been a German colony until the end of the First World War. Under the Treaty of Versailles, it was given to Australia as a League of Nations mandate. Britain had held the southern portion of the east of the island, but by this time had ceded control to Australia. The journey, as Mayr recounted it, exemplified the bravado of his youth. Against all advice, he canoed along the northern coast from the Dutch colony to the Mandated Territory. There was no overland route and he could not afford the more common route of going first to Australia.

His area of zoological research was ornithology (the study of birds), focusing specifically on the classification of different species. Classification of organisms is the domain of taxonomy (sometimes also referred to as systematics – systems for organizing organisms). The prevailing system of classification was that developed by Carl Linnaeus, which is still used today. It is known as the binomial system. All organisms are identified as belonging to a species within a genus. This is a unique identifier. Contemporary humans, for example, are *Homo sapiens*. The first term is the genus category to which we belong (hominids) and second term the species. This is a brilliant system which over time has been expanded and revised by new discoveries. It is largely a bookkeeping system, however, akin to a catalogue of the size and position of the stars and planets. Knowing what things exist, documenting their characteristics and grouping them in useful ways are essential parts of science, but these activities do not address questions about origins or the mechanisms of change. That is a second and very different step.

Mayr's work in the interior of New Guinea placed him on the path to this second step. He not only documented and classified many new birds – especially birds of paradise – but also noticed that in the various mountain peaks along the chain of mountains running through the centre of the island, the birds varied. What he was seeing was what Darwin had seen on the Galápagos Islands. The birds were similar in important respects but differed in other important traits. For Darwin it was finches and the size and shape of their beaks; for Mayr it was birds of paradise and their plumage. The mountain peaks were 'islands in the sky' (illus. 24).

Equally important to Mayr's emerging understanding of evolution was Dobzhansky's *Genetics and the Origin of Species*. The key idea that Mayr gained from it was reproductive isolation, but his approach was from a different perspective. As an ornithological taxonomist his focus was on the classification of species. He was interested in explaining the biological basis for deeming something a species and how different species came to be. Like Darwin, his work made clear that there were significant differences in traits of organisms classified within a single species. He observed gradations in these traits – small differences that gave a sense of transition. Again like Darwin, he observed that these differences had a geographic element.

These were the observations of a naturalist with a keen eye and methodical documentation, which resulted in his finding and naming more that twenty new species of birds and almost 40 orchids. This alone would have given him a place of prominence in biology. Yet again like Darwin, however, he went further and sought explanations. Darwin had argued that species were artificial classifications and that natural selection acting on small individual variations was the principal mechanism of evolution. Both arguments were widely accepted by 1940. Species from Linnaeus's time had been classified

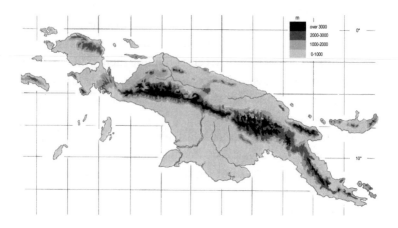

24 Map of New Guinea showing the mountain range down the middle, with shading to show the elevations.

on the basis of shared characteristics. Darwin's analysis demonstrated that this was problematic. In terms of characteristics, some sub-species within a species varied more from each other than from sub-species of a different species. Moreover, there was no evolutionary or genetic basis for the classification. It rested solely on the majority opinion of experienced naturalists: another form of stamp collecting.

Darwin had also recognized the branching pattern of evolution – the tree of life. Natural selection acting on small individual variations will continually modify populations of organisms. There is nothing to explain divergence, however. There will be 'descent with modification' but it could be entirely linear. The multiplicity of life forms requires divergence: many arising from a few. Without an evolutionary explanation of divergence, systematics was again just a catalogue of what exists.

Mayr made two important contributions in his *Systematics and the Origin of Species* of 1942, both derived from Dobzhansky. First, he offered an evolutionary/genetic basis for classification: species are groups of organisms that cannot interbreed. They either cannot produce offspring or, if they can, those offspring will be sterile, as mules are. This, rather than the number of characteristics they share, is what makes a group a species.

This definition of species has problems. It does not work for asexually reproducing plants or for certain organisms known as ring species. *Rana pipiens*, a species of frog, is such a species. Each group adjacent to each other down the Atlantic coast of the United States can interbreed, but those at the northern end of their range cannot breed with those at the southern end.

Second, he explained divergence. One species will give rise to two or more species if sub-groups are reproductively isolated for a long period of time. The most common cause of this isolation was geographic. A species could become geographically isolated in many ways: for example, the formation of a mountain chain which separates a once randomly interbreeding population, or a chance, or even deliberate, migration of a sub-group to a distant island or mountain peak. The birds Mayr studied were on different mountain peaks, and infrequently traversed the valleys between them. Once something decreases or eliminates the interbreeding between two or more sub-groups, Dobzhansky's genetic mechanism applies. That is, when two or more populations of organisms are reproductively isolated, different mutations will accumulate and eventually the two groups will be reproductively incompatible. At this point there will be two species. This, the second of Mayr's contributions (a genetically and environmentally based mechanism for species formation – a process called speciation), explains why reproductive incompatibility provides an evolutionary basis for species classification. Systematics has been brought into the evolutionary fold.

George Gaylord Simpson (1902–1984) began his university education at the University of Colorado at Boulder, where he took a geology class from Arthur Tieje. That experience was transformative. He transferred to Yale for his final undergraduate year, which had a strong reputation in geology and palaeontology. His graduation was delayed because Yale had additional requirements over Colorado, including a foreign language requirement. In 1926 he received a PhD from Yale. His doctoral supervisor was Richard Swan Lull, an excellent palaeontologist. His dissertation focused on mammals of the American west from the Mesozoic Era. Yale had an impressive collection of remains.

The Mesozoic Era ('middle animals' era) consists of three periods: the Triassic (251–199.6 million years ago), the Jurassic (199.6–145.5

million years ago) and the Cretaceous (145.5–65.5 million years ago). During the Triassic period, all the contemporary continents were joined, forming a single landmass – Pangaea – except for South China. By the Cretaceous, parts that are today continents were drifting apart (illus. 25, 26).

By the end of the Cretaceous period, a mass extinction had occurred. By some estimates, 60+ per cent of marine species and 10+ per cent of land species became extinct. The most well-known extinction is that of dinosaurs, all of which disappeared, except those in which flight had evolved. These extinctions coincide with the deposition of a layer of clay containing high levels of iridium. Iridium is found in the core of the earth and in extraterrestrial bodies, leading many researchers to hypothesize that either a massive impact of an extraterrestrial body or extreme volcanic activity occurred during this period.

Simpson and Lull had a rocky relationship, but Simpson's doctoral studies gave him an excellent grounding in existing palaeontological knowledge as well as research methods. He continued his research on primitive mammals at the Natural History Museum in London and returned to the u.s. late in 1927 to take up the position of Assistant Curator of Vertebrate Paleontology at the American Museum of Natural History. His work illuminated two aspects central to evolution. The first was that Darwin was correct about the branching pattern of evolution. Simpson's work on the evolution of the horse is a classic example. He showed a branching pattern contrary to the prevailing view that the evolution of the horse was linear. The importance of this went well beyond showing that Darwin's view was correct. It also demonstrated that the only explanation of this branching pattern was an evolution to different geographic regions as the populations spread and occupied different habitats. The different habitats gave rise to different selective pressures. As Mayr had argued, populations that become geographically separate have different selective pressures and therefore follow different evolutionary pathways. Simpson had shown that palaeontology provides confirming evidence of Darwin's theory and that Darwin's theory explains the discoveries of palaeontologists. Palaeontology had been brought into the synthesis.

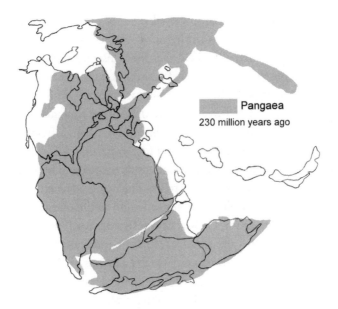

25 The single continent Pangaea *c*. 230 million years ago.

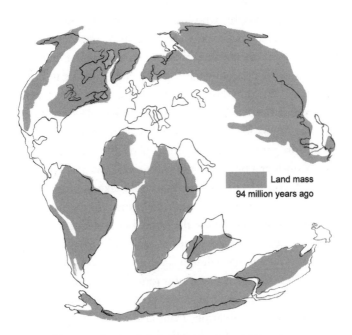

26 Pangaea broken into continents separated by oceans
c. 94 million years ago.

Second, and quite controversially, Simpson met the problem of apparent discontinuities in the palaeontological record head on. Religious fundamentalists held that these discontinuities disproved evolution. Darwin had explained them in terms of an imperfect fossil record. Fossils are only preserved under special circumstances (sedimentation, pressure, volcanic deposition and so on). Many palaeontologists, however, were inclined to see them as evidence that natural selection acting on individual variations was false. Most believed that evolution required saltations (major mutations). Fisher had shown that Mendelian heredity did not reduce variation and that it was compatible – theoretically and mathematically – with Darwinian gradualism. He had demonstrated that Darwinian gradualism was theoretically possible. He did not demonstrate that that was how things 'actually evolved'. Many, perhaps most palaeontologists in the 1930s held that the evidence suggested that even if gradualism were theoretically possible, it was not the way in which evolution had occurred.

Simpson drew on the work of another founder of population genetics, Sewall Wright, who had championed the idea of random genetic drift as an important mechanism of population change. Simpson saw in this mechanism a way to explain the apparent discontinuities in the palaeontological record while adhering to Darwinian gradualism. Wright's concept of genetic drift is complicated. Understanding the details requires significant knowledge of probability theory, but the basic idea is reasonably easy to grasp.

Random genetic drift occurs as a result of sampling during mating. Consider a bin containing 5,000 cashews and 5,000 pecans, from which you select two nuts. You could take out two cashews or two pecans, or one of each. If you replace the nuts and sample again, the possible combinations are the same, but you might get a different combination. If you are sampling randomly, over many samples, say 10,000, you will have drawn from the bin many different combinations. Let us scale it down a bit to make it manageable. Instead of drawing out two nuts, imagine you select a sample of 2,000. You could draw any combination, from all cashews to all pecans, and every mix in between. An old friend from the Galton and Pearson era describes this array of possibilities. The binomial and normal distributions describe the likelihood (probability) of the various combinations. Remember that the binomial distribution and the normal distribution are nearly indistinguishable after about 1,000 trials. So over many samples, the most common combination will be 1,000 cashews and 1,000 pecans. All cashews and all pecans are out on the tails of the distribution and are highly improbable but not impossible. We have to assume an equal number of cashews and pecans but the bin need not contain an equal number of cashews and pecans. It could contain 7,000 cashews and 3,000 pecans. A sample of 2,000 would still cover the range of combinations. Each sample has a high probability of yielding a different combination.

Instead of cashews and pecans, random genetic drift uses alleles, *A* and *a* for example, at a locus. The bin is a population: a group of randomly interbreeding individuals. The 'selecting' from the bin is the mating of individuals. This in probability is called sampling. From all the gametes (egg and sperm), a certain number will combine during mating. Mating in a population is continuous, but for mathematical purposes, it can be considered discrete with small temporal steps. One might choose periods of 24 hours. Consequently, the population is sampled 365 times in a year. The smaller the temporal period, ten minutes for example, the larger the number of samples over a year or generation.

What mathematical simulations demonstrate is that in very large populations, over a large number of mating periods – a large number of samplings – the most frequent combinations of *A* and *a* are within 10 per cent of the proportion in the whole population. So if the whole population has 0.6 *A*s and 0.4 *a*s, the most frequent proportions in mating samples will be between 0.54 *A*s to 0.46 *a*s and 0.66 *A*s to 0.34 *a*s. Any mating sample that differs in the proportion of *A*s to *a*s from the proportion in the whole population is drift. The drift from the whole population proportions is non-directional, unlike natural selection. That is, in one mating period the mating sample could be 0.54 *A*s to 0.46 *a*s and in the very next one it could be 0.66 *A*s to 0.34 *a*s. The proportion of genes drifts around but in no particular pattern.

Although drift is non-directional, it might by chance be the case that 300 temporally sequential mating samples all differ from the whole population in one direction – continuously more *A*s than *a*s for example. In that case, even in very large populations, drift will have evolutionary consequences. Sometimes drift alone will change the genetic composition of a population. Any small change in gene frequency is gradual evolution.

Drift becomes even more significant in populations that are not very large. Allelic proportions in small mating samples can deviate significantly from the proportion in the whole population. Different samples will differ from each other in allelic proportions, and different populations that had similar allelic proportions at some point in time will diverge as a result of drift. Sometimes a population becomes very small, either through a collapse of the population or through a small group becoming separated from the main population. Such small groups often fail to survive but sometimes they do and the effect of drift is dramatic, resulting in a new species very quickly.

Simpson saw in this phenomenon an explanation of how there can be long periods with few new species and then a burst of new species appearing. Drift, along with selection acting on small variations, will lead to this episodic pattern. Consequently Simpson eroded the basis on which the palaeontological community rejected Darwinian gradualism. His two contributions – the branching pattern of evolution, exemplified by the horse, and the role of drift in creating an episodic fossil record – brought palaeontology into the Darwinian/Mendelian synthesis.

Plants are incredibly diverse genetically, developmentally, reproductively and phenotypically. Their genetic and reproductive diversity was a major challenge for evolutionary theory. Some plants only reproduce asexually. A sizeable number of these are self-fertilizing (hermaphrodites). Many others reproduce by budding, branching, producing a side shoot at ground level (tillering) or producing spores or seeds that are genetically identical to the parent plant. Some plants reproduce both asexually and sexually. Some of these, vascular plants for example (plants that have a system of fluid-filled vesicles similar to the blood system in animals), have sexual cycles and asexual cycles as sub-cycles of a larger cycle. Some plants reproduce only sexually. The evolutionary theory of Fisher, Haldane and Wright focuses on sexual reproduction. Many species of plants fall outside its scope.

Moreover, many plants are, or become during reproduction, polypoid (having three or more matched chromosomes). Cells of most animals are diploid (their chromosomes occur in pairs). Sperm and egg cells in sexually reproducing organisms are haploid (their chromosomes have no partner) (illus. 27). Having three or more partnered chromosomes is very rare in animal cells, though in plants it is widespread (illus. 28). It took until 1950 for these features of plants to be incorporated into the evolutionary consensus. A major player in this was G. Ledyard Stebbins.

G. Ledyard Stebbins met Edgar T. Wherry, professor of botany at the University of Pennsylvania, in the summer of 1925 after his freshman year at Harvard. Stebbins was studying plants in the region of Bar Harbor, Maine. Wherry was a specialist in ferns. That summer was a watershed. As a freshman, Stebbins was unable to decide on an area of study. After that summer he was committed to a career in botany. He completed his PhD at Harvard in 1931, where he was exposed to the cytogenetic study of plants championed by Karl Sax. After a brief stint at Colgate University, he was offered a research position at the University of California, Berkeley, by Professor Ernest Brown Babcock. It was during his time there that he did research on polyploidy. He developed a close association and friendship with

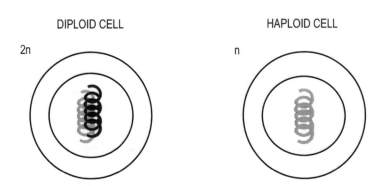

DIPLOID CELL   HAPLOID CELL

2n   n

27 A diploid cell has two sets of chromosomes; this is the norm for animal cells. Each chromosome from one set has a companion in the other set. In a resting phase, the matched pairs are joined. During cell division, the matched chromosomes separate. For each chromosome a new matching chromosome is built, resulting in four sets of chromosomes. When the cell divides, two matched sets go into one cell and the other matched set into the other cell. A haploid cell has only one set of chromosomes. Gametes (sex cells) are haploid.

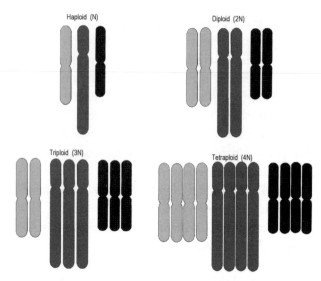

28 Haploid cells (sex cells, for example) have only one of each chromosome (top left). Cells in animals are usually diploid (top right) – they have paired chromosomes. Polyploid cells have more than two associated chromosomes (bottom left and right).

Dobzhansky. This resulted in his giving the Jesup Lectures at Columbia in 1947. These lectures formed the basis for his ground-breaking book of 1950, *Variation and Evolution in Plants*. It provided a wealth of analysis and explanation of plant genetics. He explained the evolution of plant species, including the tempo (speed) of plant evolution. His work on tempo made sense of the fossil record.

In the fifth chapter he provides evolutionary genetic explanations of self-fertilization, the origin of the diploid state and the differences and similarities between the genetic systems in plants and animals. In chapter Six, he examines the role of isolation in the origin of plant species, connecting the origin of plant species to the work already done by Dobzhansky, Ford, Mayr and Simpson. Their work was mostly on animals. In chapters Eight and Nine, he examines and explains polyploidy from an evolutionary perspective. In chapter Fourteen, he tackles fossils. He looks at the modern distribution patterns and rates of evolution. He thereby completed in botany the work on these issues started by Simpson, who focused on animals. The volume is a tour de force and in one sweep brings the plant world into the modern synthesis.

Like the work of Dobzhansky, Ford, Mayr and Simpson, Stebbins's work was enriched and modified from 1950 to the present. In the early years after *Variation and Evolution in Plants*, it was Stebbins who contributed to a more sophisticated knowledge and revised explanations. Since then, however, all botanical research has only increased the explanatory power of the Darwinian-Mendelian theory. The genetics of animals and plants are now known to be complex, and simple Mendelian inheritance does not capture that complexity. For instance, as biometricians have noted, many traits are quantitative. Different organisms within a species will have different values – height, number of hair follicles and quantity of milk production, for example. Today population genetics embraces remarkably complex systems, further attesting to the robustness of the theory of evolution. The modern synthesis was the fulfilment of Darwin's quest for a consilience of inductions.

At this point, microbiology (archaea, bacteria, fungi, protista and viruses) remained outside the evolutionary fold. It is an oddity of the development of the modern synthesis that S. E. Luria and M. Delbrück's landmark work was not mentioned by those who forged the modern synthesis, with, as noted later, Dobzhansky being the only exception. Their 1943 article in *Genetics* demonstrated with statistical analysis and experimental findings that bacterial resistance to viruses was due to individual, pre-existing, heritable variability in bacteria and that Darwinian selection was at work in increasing the bacteria that were resistant. They were awarded the Nobel Prize in Physiology and Medicine for this work in 1969.

Their demonstration involved consideration of two competing hypotheses:

1) *First hypothesis (mutation):* There is a finite probability for any bacterium to mutate during its life time from 'sensitive' to 'resistant'. Every offspring of such a mutant will be resistant, unless reverse mutation occurs. The term 'resistant' means here that the bacterium will not be killed if exposed to virus, and the possibility of its interaction with virus is left open.

2) *Second hypothesis (acquired hereditary immunity):* There is a small finite probability for any bacterium to survive an attack by the virus. Survival of an infection confers immunity not only to the individual but also to its offspring. The probability of survival in the first instance does not run in clones. If we find that a bacterium survives an attack, we cannot from this information infer that close relatives of it, other than descendants, are likely to survive the attack.

The last statement contains the essential difference between the two hypotheses. On the mutation hypothesis, the mutation to resistance may occur any time prior to the addition of virus. The culture therefore will contain 'clones of resistant bacteria' of various sizes, whereas on the hypothesis of acquired immunity the bacteria that survive an attack by the virus will be a random sample of the culture.

The first is an evolutionary hypothesis; the second is quasi-Lamarckian and not consistent with the emerging evolutionary genetics perspective. The key sentence in their paper is (emphasis added):

It will be seen that in every experiment the fluctuation of the numbers of resistant bacteria is tremendously higher than could be accounted for by the sampling errors, in striking contrast to the results of plating from the same culture . . . *and in conflict with the expectations from the hypothesis of acquired immunity.*

This should have begun the process of bringing microbiology into the evolutionary fold. Yet among those who forged the modern synthesis, only Dobzhansky, in the 1951 edition of *Genetics and the Origin of Species*, mentioned Luria's and Delbrück's work, claiming it to be a 'brilliant analysis', with explicit to attention paid to the issue of Lamarckianism. He also mentions similar experiments by Demerec and Fano (1945) and Luria (1946). It therefore appears that some geneticists knew about this work.

Jepsen, Mayr and Simpson published *Genetics, Paleontology and Evolution* in 1949; this was an exceptionally important contribution to the institutionalization of the *Modern Synthesis*. It does not mention

Luria and Delbrück. There is nothing on *E. coli* and only one brief mention of bacteria focused on their rate of evolution. There is not a single reference to Luria and Delbrück, or to *E. coli* or bacteria, in Huxley's and Hardy's edited volume *Evolution as a Process* (1954). Ernst Mayr's collected essays, *Evolution and the Diversity of Life* (containing almost all the important papers by Mayr), does not have a single reference to Luria and Delbrück. Nor does the retrospective volume *The Evolutionary Synthesis* (1980), edited by Mayr and Provine. The same is true of G. G. Simpson's *The Meaning of Evolution* – not even in the 1960 edition. In 1960 Sol Tax published a three-volume treatise that encompassed almost everything that had been discussed among the synthetists at that time. The only reference to Luria and Delbrück is again made by Dobzhansky. Moreover, there are almost no references to microbes. Microbiology remained on the periphery of the synthesis.

Embryology also remained on the periphery. It was an important analogue to evolution in Darwin's reasoning. J.B.S. Haldane's famous exchange, cited by Richard Dawkins in his *The Greatest Show on Earth* (2010), exemplifies its analogical value. During one of his lectures, a sceptical woman commented:

> Professor Haldane, even given the billions of years that you say were available for evolution, I simply cannot believe it is possible to go from a single cell to a complicated human body with its trillions of cells organized into bones and muscle and nerves, a heart that pumps without ceasing for decades, miles and miles of blood vessels and kidney tubules, and a brain capable of thinking and feeling.

Haldane responded: 'But Madam, you did it yourself! And it only took nine months.' Embryology's heuristic value was clear to evolutionary biologists, but its role in the dynamics of evolution was under-appreciated until the 1980s. One possible reason for this delay in incorporating microbiology and development into evolutionary dynamics was the importance of molecular biology to successful integration. That area of biology was not sufficiently developed until the 1960s (for more on this see chapter Ten).

EIGHT

Extending Genetics: The Molecular and Chemical Basis of Heredity and Life

In 1869, ten years after the publication of *The Origin of Species*, Friedrich Miescher identified a substance in the nucleus of eukaryotic cells: a substance which, he determined, was not a protein. In organic chemistry – the branch of chemistry that deals with carbon containing molecules – the suffix 'ease' attached to a compound (or its initial syllable/s) means that it breaks the named compound apart. 'Protease' breaks apart proteins. Nuclein, the name Miescher called the substance he identified, was not broken apart by protease, suggesting it was not a protein. With this identification, the path to the discovery of the structure and importance of DNA had begun.

Fred Neufeld (1869–1945), a bacteriologist in Germany, published his discovery of three strains of pneumococcal bacteria in 1902. He discovered the three strains by using antiserums, antibodies that have developed in response to specific antigens. Antigens are material foreign to an organism that triggers the immune system to develop antibodies to destroy the antigen. A successful immune system's production of antibodies will clear the blood of the antigen; an unsuccessful one results in illness and sometimes death. Once produced, antibodies stay in the blood for a long time, often forever, and provide a faster response the next time the antigen appears. This is the basis of vaccination and of intravenous immunoglobulin production. There are different kinds of vaccines: live antigens, attenuated (weakened) antigens and dead antigens. Antigens are most commonly weakened by passing them through another animal first. The goal of vaccination is to cause the immune system to produce antibodies under controlled conditions so that when the 'wild-type', more virulent (stronger) antigen is encountered the immune system is ready.

Plasma is the fluid part of blood; the solid parts are red cells, white cells and platelets. When blood clots, the fluid part is called serum. Serum (and plasma) contains the antibodies that an organism has produced. An antiserum is serum that contains antibodies to a specific antigen. Neufeld collected serum from animals that had been exposed to pneumococcal bacteria. When pneumococcal bacteria were exposed to pneumococcal antiserum, they swelled. Different pneumococcal bacteria swelled when exposed to different antiserum; that is, some did not swell with some antiserum. He was able to demonstrate that three kinds of antiserum and three kinds of pneumococcal bacteria existed. He called the swelling a Quellung reaction,

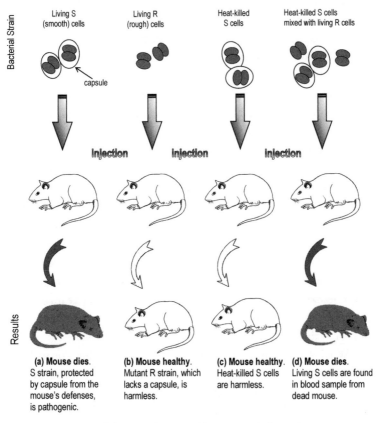

29 Griffith's experiments with pneumonia bacteria.

from the German word for swelling. Neufeld also discovered that ox bile would destroy pneumococci, which dissolve in it. This was a crucial discovery in the diagnosis of pneumenococcal infection. If the bacteria dissolved in ox bile, the diagnosis was positive.

In 1928 the British microbiologist Frederick Griffith (1877–1941) devised and performed brilliant experiments using *pneumoniae* bacterium. He worked with two strains: a virulent strain (Smooth cell *pneumoniae* bacterium: S cells), which has a protein and sugar envelope (outer casing); and a non-virulent strain (Rough cell *pneumoniae* bacterium: R cells), which is un-enveloped.

In a sequence of four experiments, Griffith discovered a process of transformation (illus. 29). He first injected living S cells into a mouse, with the expected result that the mouse died. A blood analysis, again as expected, found live S cells. The envelope protected the bacterium from successful attack by the mouse's immune system. He then injected living R cells into a mouse, with the expected result that the mouse survived. A blood analysis found no R cells. The mouse's immune system was capable of destroying them. These two experiments confirmed the different effects of the strains. Next he heat-treated S cells, which removed the envelope. He injected these into a mouse. The mouse survived and no S cells were found in the blood. This verified his hypothesis that it was the envelope that protected the S cells from the mouse's immune system. It was his fourth and final experiment that produced a surprising result. He injected a mixture of heat-treated S cells and living R cells into the mouse. The expected result based on the second and third experiment was that the mouse would live and no S cell or R cell would be found in its blood. The mouse died and enveloped, virulent S cells were found in its blood. Griffith correctly hypothesized that the R cells had been transformed into S cells. He also hypothesized that the transformation involved the nuclear material in the cells, although he did not identify it as DNA; the state of knowledge of DNA at that time made that identification impossible.

With our current knowledge of DNA, it is possible to provide an explanation of transformation in bacteria. In this case, the heat treatment of the S cells destroyed their envelopes, which released their DNA. The R cells took up that DNA. The S-cell DNA coded for an envelope, which then protected the R cells that contained R-cell DNA and S-cell DNA. This is one of the first observations of horizontal gene transfer, which occurs frequently in bacteria, and is one of the ways in which they quickly evolve resistance to antibiotics and evolve protection from the immune system or other environmental challenges. They keep swapping DNA with a high frequency, and some of the new combinations will increase fitness in a specific environment.

A next significant step in this journey drew on Griffith's experimental technique and his findings. In 1944 Oswald Avery, Colin MacLeod and Maclyn McCarty, working in the U.S., published the results of experiments using *Streptococcus pneumoniae*. Avery was the principal investigator and his name is the one associated with the work. Like Griffith, Avery used S cells and R cells. Unlike Griffith, whose experiments were *in vivo* (Latin: literally 'in the living', commonly today defined as 'in the living body of a plant or animal'), Avery's experiments were *in vitro* (Latin: literally 'in [or under] glass', commonly defined today as 'outside the body'). This required some special techniques and care. He discovered that a particular strain of R-variants (R36A) had two important properties, making it ideal for these experiments. First, it has never been observed to revert to the Type II S form and attempts to induce a reversal have failed. So it is relatively stable. Second, it transforms readily to S form in the presence of killed S cells.

The essence of Avery's experiments was to treat killed S cells in one of three ways: with protease (a protein degrading enzyme), with RNase (a RNA-degrading enzymes) and with DNase (a DNA-degrading enzyme). He then mixed each of these batches with R36A cells. The results were:

1) The protease-treated killed S cells still transformed.

2) The RNase-treated killed S cells still transformed.

3) The DNase-treated killed S cells did not transform.

The conclusion he drew was that DNA was the genetic material required for transformation. The experiments were meticulously conducted and the reasoning involved was impeccable. Moreover, the results were far-reaching, establishing that DNA was a special molecule

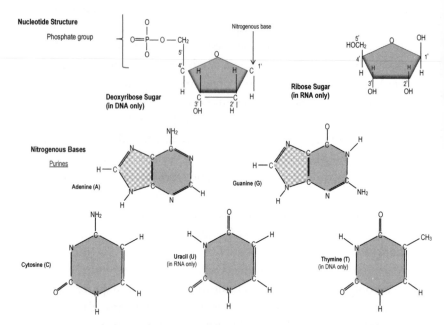

30 The chemical structure of nucleotides. The upper diagrams are of the common structure of all DNA (left) and all RNA (right). The nitrogenous bases are added where indicated. Which one is added determines which specific nucleotides are distinguished. In DNA, uracil is the base and in RNA thymine is the base.

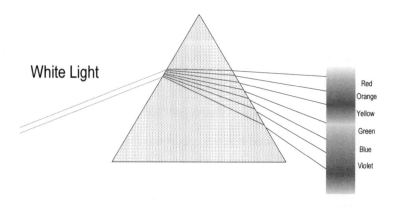

31 White light passing through a prism is broken into its component colours, represented here, in black and white, by the different bands of shading.

and probably the hereditary material. Nonetheless, he was never awarded a Nobel Prize.

The next major discovery came in 1950 when Erwin Chargaff discovered an interesting relationship. He examined DNA from numerous species and found that the proportion of nucleotides was constant and uniquely paired. A nucleotide is a component of DNA. There are four nucleotides found in DNA: adenine (A), guanine (G), cytosine (C) and thymine (T). They all have a common chemical core structure. Their differences result from different nitrogenous bases. What Chargaff discovered in the DNA of all the numerous species he examined was that the proportion of adenine to thymine was equal and constant, as was the proportion of guanine to cytosine. That is, the amount of adenine equalled the amount of thymine (A = T) and the amount of guanine equalled the amount of cytosine (G = C) (illus. 30).

An understanding of the significance of this equivalence would come three years later, when James D. Watson and Francis Crick unlocked the structure of DNA. The principal empirical data available to those trying to decode the structure of various organic molecules was X-ray crystallography. This is analogous to the behaviour of light as it passes from one medium (say, air) to another (say, water); it bends. Most people are aware that when light is shone through a prism, it breaks into all its component colours. The 'breaking into' results from the diffraction of the light by the prism. This is how we

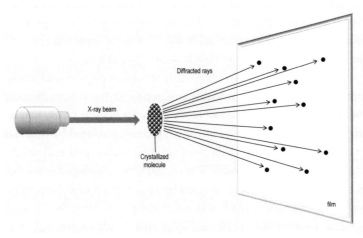

32 An X-ray beam deflected by the surface of a chemical crystal.

came to understand that white light is the combination of the colours we see (illus. 31).

X-ray crystallography works in the same way, except X-rays are being 'shone at' (directed at) the crystalline form of a molecule, rather than light at a prism. The result is a unique diffraction for each different molecule. The theoretical task involves determining what structure a molecule would have to have in order to produce that X-ray crystallographic diffraction pattern (illus. 32).

The American chemist Linus Pauling (1901–1944) was in the race to discover the structure of DNA, neck and neck with Watson and Crick. He was a brilliant biochemist whose early work was on haemoglobin, the substance in red blood cells that absorbs oxygen in the lungs and delivers it to the cell of the body. He gave a robust account of its structure. In the mid-1930s he turned his attention to the general structure of proteins.

Pauling's work on proteins was exceptionally impressive because he used the diffraction patterns produced in the 1930s by William Astbury, a British crystallographer. These were early days. Astbury's patterns were produced by diffraction off molecules that had an atypical orientation. As a result, it took Pauling more than a decade to reconcile Astbury's patterns with his mathematical analysis based on quantum mechanics. The essence of Pauling's model was that proteins have an alpha-helical structure. That is, they are shaped like the lower-case Greek letter alpha, α. In 1954 he was awarded the Nobel Prize in Chemistry for this discovery as well as his groundbreaking work on the chemical bond. This latter work was the basis for his book *The Nature of the Chemical Bond* (1939).

Pauling's brilliance in chemistry, as well as physics and mathematics, made him a formidable competitor in the race to discover the structure of DNA. James D. Watson and Francis Crick had an advantage. They used X-ray diffraction pictures produced by Rosalind Franklin and Maurice Wilkins. Franklin's techniques and resulting pictures were exceptional. Watson and Crick developed a number of models but the one that emerged by 1953 was a double helix (illus. 33). Two chemical strands wound around each other in a helical pattern. Its structure is like a ladder that has been twisted. The rungs of the ladder are composed of adenine joined to thymine or

33 Watson's and Crick's diagram as presented in their original *Nature* paper (vol. 171, 1953), with the words 'This figure is purely diagrammatic. The two ribbons symbolize the two phosphate-sugar chains, and the horizontal rods are the pairs of bases holding the chains together. The vertical line marks the fibre axis.'

guanine joined to cytosine. This structure not only explained Franklin's diffraction patterns but also Chargaff's discovery that in all DNA, A = T and G = C.

Linus Pauling had been in hot pursuit of the structure of DNA. By 1953 he had developed a triple-helical model. Watson and Crick knew that Pauling would soon realize that his model had flaws and that a double helix was a better explanatory model. There was no time to lose.

The chemical structure of DNA is easy to grasp, but the implications of its structure are complex and far-reaching. All matter is composed of atoms. Atoms are made up of three particles: protons (positively charged particles), neutrons (particles with no charge) and electrons (negatively charged particles). Their structure is like a micro solar system, with a nucleus at the centre analogous to the sun in our solar system, and electrons orbiting the nucleus analogous to the planets.

The nucleus consists of protons and neutrons. The number of electrons orbiting the nucleus is the same as the number of protons in the nucleus. The same atom, say oxygen, can have different numbers of neutrons. Atoms that have the same number of protons, and hence electrons, but differ in the number of neutrons, are isotopes of that atom. When atoms are joined together, electrons are shared.

The elements that make up nucleotides are hydrogen, oxygen, carbon, nitrogen and phosphorous. These elements have a property called valence. The valence determines how molecules can combine. Oxygen has a valence of 2. Hydrogen has a valence of 1. Hence oxygen can bond with two hydrogen elements. The product of this combination is water, H_2O. The formula indicates that two hydrogen atoms have bonded with one oxygen. Carbon has a valence of 4, nitrogen a valence of 3 and phosphorous a valence of 5. In the diagram, the short lines indicate the bonding of molecules. The double short lines indicate that two valences are occupied in the bonding (illus. 34).

34 The phosphodiester bonds are shown in boldface.

The side rails of the twisted ladder are made up of the phosphate complex of the nucleotides. They are held together by phosphodiester bonds – shown in boldface type in the diagram opposite. The two nucleotides to be joined are on the left. When they join (ligate), two hydrogen atoms and one oxygen atom are released in the form of water. The nucleotides that comprise the rungs of the twisted ladder are held together by hydrogen bonds.

A lesson on rhetoric in science can be extracted from the announcement of this discovery. In his monograph *The Double Helix* (1968), J. D. Watson recounts reading papers by Linus Pauling:

> Most of the language was above me, and so I could only get a general impression of his argument. I had no way of judging whether it made sense. The only thing I was sure of was that it was written with style . . . again the language was dazzling and full of rhetorical tricks . . .
>
> One article started with the phrase, 'Collagen is a very interesting protein'. It inspired me to compose opening lines of the paper I would write about DNA, if I solved its structure. A sentence like 'Genes are interesting to geneticists' would distinguish my way of thought from Pauling's.

Watson did exactly that when he and Crick published their abstract announcing the structure of DNA in *Nature* on 25 April 1953 – 'Molecular Structure of Nucleic Acids: A Structure for Deoxyribose Nucleic Acid'. The opening sentence reads: 'We wish to suggest a structure for the salt of deoxyribose nucleic acid (DNA). This structure has novel features which are of considerable biological interest.' This is attention-getting and also something of an understatement given the significance of their model. They then immediately cleared away alternatives to their model:

> A structure for nucleic acid has already been proposed by Pauling and Corey. They kindly made their manuscript available to us in advance of publication. Their model consists of three intertwined chains, with the phosphates near the fibre axis, and the bases on the outside. In our opinion, this

structure is unsatisfactory for two reasons: (1) We believe that the material which gives the X-ray diagrams is the salt, not the free acid. Without the acidic hydrogen atoms it is not clear what forces would hold the structure together, especially as the negatively charged phosphates near the axis will repel each other. (2) Some of the van der Waals distances appear to be too small.

After a quick dismissal of another three-chain model developed by R.D.B. (Bruce) Fraser, they moved to a concise and gripping announcement: 'We wish to put forward a radically different structure for the salt of deoxyribose nucleic acid. The structure has two helical chains each coiled round the same axis [see illus. 33].'

Calling it 'a radically different structure' suggests novelty, and bold and creative thinking. Their diagram provides a simple, elegant and stylistic presentation. It conveys quickly the general character of the structure of DNA without the complexities.

The last substantive paragraphs are teasing and a powerful understatement:

> It has not escaped our notice that the specific pairing we have postulated immediately suggests a possible copying mechanism for genetic material.
>
> Full details of the structure, including the conditions assumed in building it, together with a set of co-ordinates for the atoms, will be published elsewhere.

To announce that 'it has not escaped our notice', with no details, leaves the reader wanting more and clearly primed to seek out the next instalment. Compare this with Avery's style. His seminal article – the contents of which we have already examined – begins:

> Biologists have long attempted by chemical means to induce in higher organisms predictable and specific changes which thereafter could be transmitted in series as hereditary characters. Among microorganisms the most striking example

of inheritable and specific alterations in cell structure and function that can be experimentally induced and are reproducible under well defined and adequately controlled conditions is the transformation of specific types of Pneumococcus. This phenomenon was first described by Griffith (1) who succeeded in transforming an attenuated and non-encapsulated (R) variant derived from one specific type into fully encapsulated and virulent (S) cells of a heterologous specific type. A typical instance will suffice to illustrate the techniques originally used and serve to indicate the wide variety of transformations that are possible within the limits of this bacterial species.

On the next page, the details of the experiment are given:

Transformation of pneumococcal types *in vitro* requires that certain cultural conditions be fulfilled before it is possible to demonstrate the reaction even in the presence of a potent extract. Not only must the broth medium be optimal for growth but it must be supplemented by the addition of serum or serous fluid known to possess certain special properties. Moreover, the R variant, as will be shown later, must be in the reactive phase in which it has the capacity to respond to the transforming stimulus. For purposes of convenience these several components as combined in the transforming test will be referred to as the *reaction system*. Each constituent of this system presented problems which required clarification before it was possible to obtain consistent and reproducible results. The various components of the system will be described in the following order: (1) nutrient broth, (2) serum or serous fluid, (3) strain of R Pneumococcus, and (4) extraction, purification, and chemical nature of the transforming principle.

This is dry, somewhat turgid prose. All the information is in the article, but Avery does not mention DNA until nearly halfway through a 7,500-word paper. Watson and Crick used 900 words and in the first

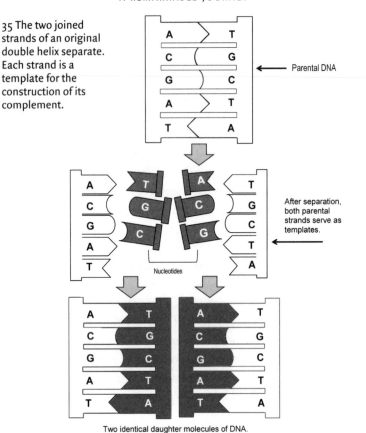

35 The two joined strands of an original double helix separate. Each strand is a template for the construction of its complement.

Parental DNA

After separation, both parental strands serve as templates.

Nucleotides

Two identical daughter molecules of DNA.

four short paragraphs provided their thesis. Avery is tentative about his conclusions. Watson and Crick are brief and make authoritative claims.

Despite the groundbreaking nature of his discovery, Avery was not awarded a Nobel Prize. Politics within the Nobel committee is probably a part of the explanation, but Andrew Bains has suggested on numerous occasions that his style of writing – the lack of rhetorical devices – is also probably a part of the explanation.

As Watson and Crick pointed out, there is an immediate implication for a copying mechanism. The nucleotide pairings are unique – A pairing only with T, and G pairing only with C. Each strand is a mirror image of the other. Hence, given either strand of the helix, the complementary strand can be constructed. If the one strand has

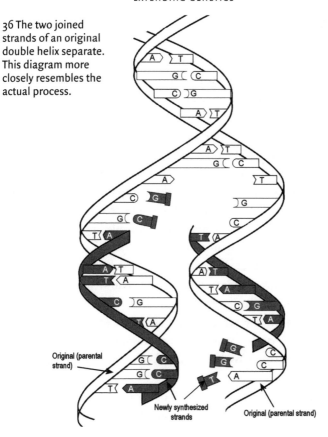

36 The two joined strands of an original double helix separate. This diagram more closely resembles the actual process.

Original (parental strand)

Newly synthesized strands

Original (parental strand)

a T, the other must have an A at that location and vice versa, and if it has a G the other strand must have C at that location and vice versa. Given a helix with the sequence AGTCG, the companion helix must be TCAGC (illus. 36, 37).

It was immediately obvious how this connects with the behaviour of chromosomes during cell division (replication). In human somatic (body) cells there are 23 pairs of chromosomes – 46 chromosomes in

----

AGTCG
TCAGC

----

37 A pattern of complementary nucleotides.

total. During mitosis (the process of cell replication) the pairs of chromosomes separate, yielding 46 separate chromosomes. For each chromosome a new mate-chromosome is created, yielding 46 pairs of chromosomes; there are two identical pairs and therefore two sets of 23, which move to the opposite ends of the cell. The cell pinches together in the middle and each half separates, yielding two cells, each with 23 pairs of chromosomes.

At the DNA level, when the pairs of chromosomes separate, the double helix separates into two strands. The single strands are perfect templates for the construction of the companion strand, as just described. Companion strands are built. The result is two complete sets of double helices. In humans, for example, 23 double helices become 46 strands. Companion strands are built for each one. This yields 46 double helices. Each helix has an identical mate: two sets of 23 helices. This is identical to the behaviour of chromosomes except at the molecular level.

At this point, the segregating behaviour of Mendel's factors (alleles), the observed behaviour of chromosomes during mitosis and the structure of DNA have been unified. They are all descriptions of the same process.

Over the next decade, the details of another marvel of DNA came into focus. Cells and their activities involve proteins. Some are the material from which cells are made (structural proteins). Others are involved in the functions that take place in the cell (functional proteins – enzymes). Hence proteins are the fundamental element of a cell's structure and how it functions.

Proteins are strings of amino acids. Twenty essential amino acids are the building blocks of proteins. We have seen how the nucleotides of DNA code for cell replication. They also code for the assembling of proteins. This is somewhat more complicated in three respects. First, more than one nucleotide is needed to code for each protein. There are twenty amino acids and only four nucleotides, so it cannot be the case that a single nucleotide codes for amino acid. Similarly, a combination of two nucleotides cannot code for twenty amino acids. There are only sixteen unique pairs of four nucleotides (4×4). A combination of three would do the job ($4 \times 4 \times 4 = 64$). That is exactly the coding that has evolved. Nature is usually parsimonious: that is, it favours the minimum necessary. Clearly 64 is many more unique combinations than are necessary, but it is the minimum required. Hence there is a lot of redundancy in the code, as the chart on p. 170 indicates.

The code for building proteins resides in DNA but there are several intermediate steps in the process from DNA to completed protein. Those steps involve another kind of nucleic acid, RNA (ribonucleic acid); there are different kinds and each has a separate role. A description of just the first step will suffice to indicate how much more complicated coding for and building proteins is compared to cell replication. One of the RNAs is known as messenger RNA (mRNA). The production of mRNA is similar to DNA replication. A strand of DNA is copied. It is the resulting mRNA that is 'read' to know which amino acids to connect to which amino acids.

What makes RNA different from DNA is that there is a substitution of a different nucleotide for thymine. Wherever thymine would be inserted in DNA replication, uracil is inserted. This process is called transcription. DNA is transcribed into mRNA. Reading the code of mRNA is called translation. The code of mRNA is translated into a string of amino acids. The tabular display of the code is called the codon dictionary (illus. 38).

A sub-unit travels along a strand of mRNA. Each triplet is read and the sub-unit attaches the appropriate amino acid to the previous one, except of course at the very beginning, which is always methionine. This process is a bit like reading. Eyes move from word to word and the mind builds a sentence by connecting the words together into a whole. Words vary in length (from one to very long strings). Triplets of nucleotides are all one length (three letters).

A quick glance at proteins reveals a fascinating world. Proteins are not linear (straight). They are folded in many complex ways and the folding changes their properties. This can be illustrated by considering bovine spongiform encephalopathy (BSE), a disease of the neural system more commonly known as 'mad cow disease'. It is called 'spongiform' because when the brains of animals, including humans, that have the disease are examined they exhibit holes similar to the appearance of a sponge. The most widely accepted explanation of the disease is that proteins produced in the brain become abnormal and infectious. They have abnormal folding. Stanley Prusner, in 1982, coined the term 'prion' for this protein. Prusner called it a protinacious (pro) infectious (in) protein. (This, of course, would make the acronym 'proin'. Prusner thought it flowed better as 'prion'.) Bovine spongiform encephalopathy is only one of many transmissible spongiform encephalopathies (TSEs). Scrapie in sheep, transmissible mink encephalopathy, feline (cat) spongiform encephalopathy and Creutzfeldt-Jakob Disease (CJD) in humans are other examples. Two key features of BSE made it especially worrying. First, all previous known TSEs appeared to be species-specific; in other words, they cannot be transmitted to other species. BSE appears to be transmissible to humans. Transmission of other TSEs requires neurological to neurological contact. BSE appears to be transmissible gastro-intestinally. Eating an infected animal is a route of transmission.

Once an abnormal prion protein is in neurological tissue or fluid, it begins to convert other 'normal' proteins into its folding pattern. Much is now known about a special class of proteins called chaperone proteins. These fold proteins into their pattern. Usually these chaperone proteins fold proteins into 'normal' non-pathogenic structures.

Second Position

		U	C	A	G	
U		UUU ⎫ Phe UUC ⎭ UUA ⎫ Leu UUG ⎭	UCU ⎫ UCC ⎬ Ser UCA ⎪ UCG ⎭	UAU ⎫ Tyr UAC ⎭ UAA Stop UAG Stop	UGU ⎫ Cys UGC ⎭ UGA Stop UGG Trp	U C A G
C		CUU ⎫ CUC ⎬ Leu CUA ⎪ CUG ⎭	CCU ⎫ CCC ⎬ Pro CCA ⎪ CCG ⎭	CAU ⎫ His CAC ⎭ CAA ⎫ Gln CAG ⎭	CGU ⎫ CGC ⎬ Arg CGA ⎪ CGG ⎭	U C A G
A		AUU ⎫ AUC ⎬ Ile AUA ⎪ AUG Met	ACU ⎫ ACC ⎬ Thr ACA ⎪ ACG ⎭	AAU ⎫ Asn AAC ⎭ AAA ⎫ Lys AAG ⎭	AGU ⎫ Ser AGC ⎭ AGA ⎫ Arg AGG ⎭	U C A G
G		GUU ⎫ GUC ⎬ Val GUA ⎪ GUG ⎭	GCU ⎫ GCC ⎬ Ala GCA ⎪ GCG ⎭	GAU ⎫ Asp GAC ⎭ GAA ⎫ Glu GAG ⎭	GGU ⎫ GGC ⎬ Gly GGA ⎪ GGG ⎭	U C A G

First Position (left) · Third Position (right)

38 The codon dictionary: the triplets of U, C, A and T specify an amino acid – as indicated. There is redundancy; each amino acid has more than one code except for methionine. Methionine always starts the building of a protein; UAA, UAG and UGA signal the end of building the protein.

Abnormal prion proteins fold 'normal' ones into their own abnormal structure, making them pathogenic. This illustrates how significant the folding pattern is. It is not just the order of amino acids that makes proteins different, but also their folding pattern (called conformation).

The explosion of knowledge within protein chemistry from 1960 onwards has underscored time and again the importance of this second feature of DNA, namely its ability to code for the building of proteins. As mentioned, at this point in our journey, the segregating behaviour of Mendel's factors (alleles), the observed behaviour of chromosomes during mitosis and meiosis, and the molecular structure and replicating process of DNA have been unified. They are all descriptions of the same process. Moreover, the role of DNA in coding the structure and functions of all cells has resolved the mystery of why cells have their individual properties. The relevance of these discoveries to evolutionary theory is clear. Mendel's theory stands at the core of evolutionary theory, but both it and the elaborations of it by Fisher, Haldane and Wright were mathematical. The discovery of chromosomes provided a physical basis for what Mendel had postulated. The discovery of the structure and behaviour of DNA provides compelling support for Mendel's theory. But it makes yet another major contribution to evolutionary theory, shedding additional light on the problem that vexed evolutionists in the 60 years after the publication of *The Origin*: variation.

Research in the 1960s using a technique called gel electrophoresis revealed that at the molecular level there was incredible variation within populations. There was often more variation within populations than between them. It would therefore take a very long time for natural selection to deplete variation. Moreover, crossing over, which had been observed by Morgan at the level of chromosomes (see p. 133, illus. 22), is now explainable in molecular terms. The chart opposite shows the process at the molecular level. Enzymes are proteins that perform functions in cells. Some are known as restriction enzymes. They break (cleave) DNA at specific regions. There are other enzymes that ligate (join) strands of DNA together. Crossing over occurs when an enzyme cleaves DNA and initiates the process shown in the diagram overleaf (illus. 39). The result is two new double strands of DNA. This, as Morgan observed, is a rich source of variation.

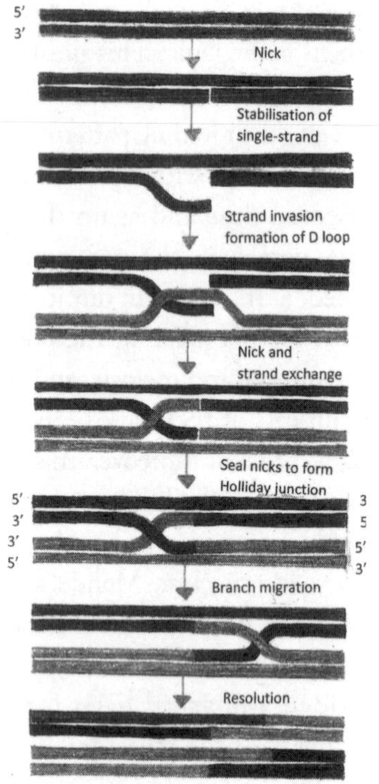

39 Strands of DNA crossing over.

DNA also resolved the mystery of mutation as a source of variation. There are four types of mutations: silent mutations, missense mutations, nonsense mutations and frameshift mutations. All are point mutations because they occur at specific nucleotide points on the strand of DNA. These four are illustrated opposite. The 'wild type' is an unmodified stand of DNA – that is, the normal and properly functioning strand prior to a mutation. Point mutations alter this wild type in the ways described. When a nucleotide substitutes for the existing wild type but still codes for the same amino acid, the mutation is silent. Nonetheless, variation has been introduced that may later be important. When a nucleotide substitutes for the existing wild type, resulting in a different amino acid code, missense mutation has occurred. This results in a new protein being produced. This may be deleterious, favourable or neutral. In the last two cases, the mutation has introduced variation that might be exploited under different selective pressure in a new environment. When a nucleotide substitutes for the

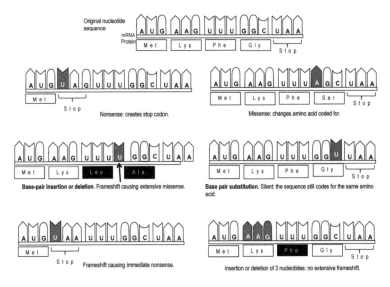

40 Different possibilities of points of mutation.

existing wild type and creates a stop codon where there was a code for an amino acid in the wild type, it is a nonsense mutation. The protein will be defective. There are two frameshift mutations. One produces missense, the other nonsense. There is also a mutation involving the deletion or insertion of a complete codon. These can be deleterious, beneficial or neutral (illus. 40).

Our knowledge of the structure and functioning of DNA has greatly enhanced our knowledge of cell replication, the way in which cells are built and what makes different cells of an organism different. These are essential elements of evolutionary theory. Moreover, it has resolved the mystery of the sources of variation, which is a crucial element of evolution because natural selection requires a continual supply of variability.

There is, however, another key to understanding why different cells in an organism are different (liver cells, brain cells and so on). After all, the organism begins with one cell. But at a point in embryological development cells differentiate, taking on different properties. That other dimension is provided by connecting embryology to evolution, making it much more than a useful analogy.

NINE

The Evolution of Behaviour: Sociobiology

Using evolutionary theory to explain the evolution of physical traits was widely accepted by the 1960s, but using it to explain behavioural traits lagged far behind. Darwin himself was fully aware that many behavioural traits have an evolutionary explanation, as many others have been since. The specifics of these explanations remained less obvious. By the 1960s it was clear that much of the behaviour of all organisms, including humans, had evolved. The most obvious examples were the behaviours of insects, especially hymenopterans (those in the order hymenoptera, which includes bees, ants and wasps). Honeybees provide a rich example because they are social insects. They cooperate for the benefit of all members of a hive. The crucial question was how a trait of a collective behaviour, such as cooperation, could have evolved.

In the early 1960s many biologists held that the evolution of social behaviour required that selection occur at the group level. This is in contrast to physical traits, which for the most part involve selection of individuals, not groups of individuals. In 1962 V. C. Wynne-Edwards published his influential *Animal Dispersion in Relation to Social Behavior*, in which he articulated a clear group selection account of social behaviour. Four years later, in 1966, George C. Williams published an equally influential book, *Adaptation and Natural Selection: A Critique of Some Current Evolutionary Thought*. In it he took apart the group selection of Wynne-Edwards and his followers and advocated that selection acted on individuals, not groups – so a successful explanation of the evolution of a behaviour required selection acting on individuals. This requirement extended to social behaviours as well.

At the heart of the case for complete individual selection is the fact that the inherited entity is the gene. If selection does not have an impact on the genes, the trait selected will not be passed to the next generation and evolution will not occur. How could the selection of group behaviour have evolutionary consequences? Ultimately evolutionarily effective selection must select for traits of individuals because it is the genes of individuals that are inherited, not the genes of groups. Indeed, there are no genes of groups, only of the individuals in them.

Something significant occurred between Wynne-Edwards's book of 1962 and Williams's book of 1966. A major article in two parts ('The Genetical Evolution of Social Behaviour I, and II') was published in the *Journal of Theoretical Biology* in 1964. Its author, the British biologist W. D. Hamilton, had transformed the biological landscape and enriched population genetics by adding to the theory a coefficient of relatedness. The essential idea is simple. Siblings share genes, as do cousins and the like. The more genes individuals have in common, the greater the evolutionary payoff of cooperating. One's own fitness – that is, the number of one's own genes that make it into the next generation – includes the fitness of close relatives. Fitness is inclusive of relatives, so the term 'inclusive fitness' is used for this expansion of the scope of one's own fitness. For obvious reasons, this is also called kin selection. To put it starkly, if you have three siblings and they all have three offspring, more of your genes will make it into the generation than if you have one offspring and they have none (illus. 41).

As David C. Queller noted in his contribution to the 'Hamilton Symposium' (Hamilton died on 7 March 2000; papers from the Symposium were published in *Behavioural Ecology* in 2001):

> Hamilton didn't completely invent the idea of kin selection. The idea was foreshadowed by Haldane (1955), Fisher (1958), and Williams (Williams and Williams, 1957). However, none of them developed it in any detail, perhaps because they did not appreciate its general importance in nature. For that we can thank animal behaviorists, particularly those like Wynne-Edwards (1962) and Emerson (1960), who believed that cooperation was very common in nature.

Types of Social Interactions

The 'actor' in any social interaction affects the recipient of the action as well as himself. The costs and benefits of interactions are measured in units of surviving offspring (fitness).

	Actor benefits	Actor is harmed
Recipient benefits	Co-operative	Altruistic
Recipient is harmed	Selfish	Spiteful

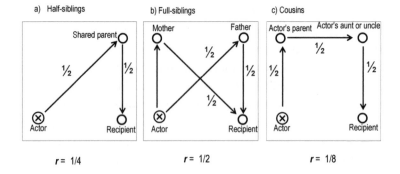

a) Half-siblings   b) Full-siblings   c) Cousins

$r = 1/4$ $r = 1/2$ $r = 1/8$

41 The upper diagram shows the combinations of outcome when two organisms interact. The lower diagram shows how relationship matters to genes being transmitted to the next generation. *r* is the proportion of genes, on average, that are shared.

Hamilton's coefficient of relatedness allowed many instances of social cooperation and altruism to be explained in terms of individual selection. Honeybees provide a perfect example. The genetics structure of honeybees is different from the genetic structure of humans. The queen and the workers (all female) are diploid: they have 16 *paired* chromosomes = 32 in total. The drones (all male) are haploid: they have 16 *single* chromosomes. There is only one queen in a hive, except during supersedure (the replacement of an old queen with a new one) or preparations for swarming (colony division resulting in two colonies, each with a queen). Normally the queen is the only bee in the hive that lays eggs. A pheromone (a chemical) produced by the queen suppresses the reproductive ability of the workers. Within a week of hatching, the queen mates with a number of drones (figures vary, but usually no fewer than seven and no more twenty) in flight. Once filled with sperm of the drones, she can either fertilize an egg, producing a worker, or not, producing a drone.

The genetics of the colony means that, on average, every worker shares 75 per cent of her genes with her sisters. As a result, a worker puts more of her genes into the next generation by helping her mother to have offspring than she would by having her own off-spring. They would have 50 per cent of her genes on average. The calculation of 75 per cent relatedness is not completely straightfor-ward. That 75 per cent relatedness is within a subfamily – all the workers that have the same father. Since there could be as many as twenty drones that mate with a queen, there could be twenty sub-families, but that would be the exception, with the norm being closer to ten. Assuming that ten drones mate with the queen, and there are 60,000 workers in the hive (the range is 20,000–80,000), on average, there will be 5,999 workers that are 75 per cent related to any specific worker. As noted, every worker has 50 per cent of the queen's genes, hence they are 50 per cent related to the queen. Moreover, one in every ten new workers will be 75 per cent related to the spe-cific worker. It is also the case that 50 per cent of the workers in the other subfamilies will share 50 per cent of her genes. The diagram over-leaf illustrates this point; it assumes that four drones mate with the queen, so there are four subfamilies (illus. 42).

The evolutionary advantages of cooperation in this case are clear, as is the case for altruism: a worker giving up her life to save the hive puts more of her genes into the next generation than being selfish. In diploid genetic systems, where both parents have a full complement of paired chromosomes, the situation is less dramatic but related-ness still matters in determining the best evolutionary strategy. As indicated, you are 50 per cent related to your siblings and 12.5 per cent to cousins (see illus. 41). That means that every offspring your siblings produce puts 25 per cent of your genes into the next gener-ation. There are many circumstances in which sacrificing your own reproduction to allow siblings to reproduce works to your genetic

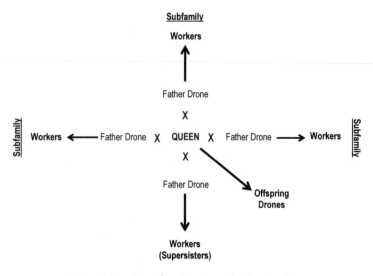

42 The relationship of workers to each other in a hive.

evolutionary advantage. Suppose that you have three siblings, both older than you, and that they already have three children each. In terms of your genes getting into the next generation, that is equal to you having 4.5 children. This is a result of the fact that, on average, ¼ of the genes of each of your nephews and nieces will be the same as yours (¼ × 9 = 2.25). When you have children, ½ of their genes will be the same as yours. Hence, to transfer 2.25 of your genes into the next generation, you will have to have 4.5 children (4.5 × ½ = 2.25). You are at a family gathering. You are outside the kitchen door. You see the house is on fire with everyone else inside. They are in the living room and the fire appears to be in the kitchen. The only way to alert them in time for them to escape through the front door is for you to immediately enter the kitchen door and start shouting, but you risk dying of smoke inhalation. From an evolutionary perspective, your altruism enhances your fitness, especially if having 4.5 offspring is unlikely. This demonstrates why the propensity to be altruistic towards siblings and their offspring has an evolutionary payoff and hence why it would have evolved. The same is true of alarm calls sounded by birds, monkeys and so on to announce the presence of predators. The alarm call makes evolutionary sense, even though it puts the alarm-sounder in danger.

Williams's arguments and Hamilton's inclusive fitness persuaded most biologists that social cooperation and altruistic behaviour can be explained using selection on individuals. Individual selection became entrenched. Some biologists continue to develop models of group selection. David Sloan Wilson is one of the most prominent of these today.

The views of Williams and Hamilton with respect to explaining the evolution of behaviour created only minor controversies. All that changed in 1975 when Edward O. Wilson's *Sociobiology: The New Synthesis* was published. Wilson is an entomologist – one who studies insects – but *Sociobiology* is wide-ranging, covering all species, including humans. This is what sparked most of the controversy. The first chapter is a discourse on the evolutionary basis of ethics and the last one is on the evolution of human behaviours. Some of the criticism was appropriate. The vitriolic nature of a significant amount of it was not.

Two vocal opponents of the views in *Sociobiology* were Stephen Jay Gould and Richard Lewontin, both colleagues of Wilson's at Harvard. Gould is principally a palaeontologist and did brilliant work on the fossils of the Burgess Shale in British Columbia, Canada. He also championed a view known as punctuated equilibria. Niles Eldridge very likely conceived the view. The first paper on it in *Paleobiology* in 1972 was jointly authored. Essentially they argued that the fossil record manifests discontinuities – long periods with little to no speciation occurring (equilibria) and then bursts during which numerous new species appear (punctuation of the stasis). We have seen this debate before. The paper attracted little attention. In 1978 they published the idea in a slightly revised form. Still the view garnered little attention. Then, in 1980 Gould, who was an excellent rhetorician and whose prose was a delight, upped the ante. In the journal *Paleobiology*, he published 'Is a New and General Theory of Evolution Emerging?' In a provocative paragraph he declared:

> I well remember how the synthetic theory beguiled me with its unifying power when I was a graduate student in the mid-1960s. Since then I have been watching it slowly unravel as

a universal description of evolution. The molecular assault came first, followed quickly by renewed attention to unorthodox theories of speciation and by challenges at the level of macroevolution itself. I have been reluctant to admit it – since beguiling is often forever – but if Mayr's characterization of the synthetic theory is accurate, then that theory, as a general proposition, is effectively dead, despite its persistence as textbook orthodoxy.

Despite its provocative closing sentence, this is a carefully crafted position – something creationists missed entirely. Gould was and remained a staunch evolutionist. He was dismissing a specific characterization of the theory. Indeed, he is careful to make explicit that it is the particular version of the synthetic theory characterized by another eminent Harvard biologist, Ernst Mayr, that is 'dead'. Mayr, as we have seen, was one of the architects of the modern synthesis, but by 1980 his characterization had been significantly modified by the increasing value of molecular genetics to evolutionary theorizing, and, as we shall see in the next chapter, the importance of embryological development. Gould is strident in this passage but the article is mostly a well-crafted articulation of the position he and Eldredge had been espousing since 1972. By 1982, for a variety of reasons (among which, one suspects, was the constant use of this passage by creationists and the extensive criticism of his complete dismissal of the synthetic theory by biologists and others – including me), Gould softened his position. The synthesis was not 'dead' but in need of significant expansion. His summary in 'Darwinism and the Expansion of Evolutionary Theory', which appeared in the widely read journal *Science* (1982), has a conciliatory tone:

> The essence of Darwinism lies in the claim that natural selection is a creative force, and in the reductionist assertion that selection upon individual organisms is the locus of evolutionary change. Critiques of adaptationism and gradualism call in doubt the traditional consequences of the argument for creativity, while a concept of hierarchy, with selection acting upon such higher-level 'individuals' as demes and species,

challenges the reductionist claim. An expanded hierarchical theory would not be Darwinism, as strictly defined, but would capture, in abstract form, the fundamental feature of Darwin's vision – direction of evolution by selection at each level.

Essentially Gould wanted to make space in evolutionary theory for punctuated equilibrium. Darwin proposed gradualism over saltations – major changes in an organism in a single generation. Gradualism is the antithesis of punctuated equilibrium; or, at least, that is how Gould saw it. The expansion for which he was looking resided in 'selection acting upon such higher-level "individuals" as demes (interbreeding populations) and species'; in other words, some form of group selection. E. O. Wilson's application of the synthetic theory, which fully embraced Darwinian gradualism, was the intellectual target of Gould's critique of *Sociobiology*.

There was also a political critique of sociobiology – especially when applied to humans. Richard Lewontin, in the 1960s, used a new technique for separating molecules of different sizes: gel electrophoresis. The details of the process are unimportant, but he used it to separate DNA of different lengths as well as enzymes. One of his findings was that there was greater genetic diversity within a racial group than between racial groups, which called into question the viability of racial and ethnic classifications. There is lots of variability within populations – lots of polymorphism (poly = many; morph = body). Lewontin is a self-proclaimed Marxist, as was Gould, for that matter. In *The Dialectical Biologist* (1985), he and Richard Levins provide an account of a Marxian dialectical approach to biology. This ideological commitment, coupled with his biological work on polymorphism, led to a rejection of Wilson's sociobiological explanations. Wilson was seen by Lewontin, Stephen J. Gould and others to be advocating a simplistic account of human behaviour that reinforced stereotypes – racial ones, among others. They maintained that Wilson had made these stereotypes appear to have a biological basis when in fact his biology was deeply flawed.

One obvious flaw, with respect to humans at least, was a lack of attention to cognition. Humans, like many other species, have acquired propensities to behave in certain ways that are evolutionarily advantageous. The propensities to have sex, to want children and to fight or flee are all part of our evolutionary heritage. They have served us well in the past. Despite their almost instinctive nature, they are only propensities. Unlike reflex actions (for example, a tap on the knee in the right place causes the leg to straighten), which are very difficult or impossible to control, most behavioural propensities can to greater and lesser extents be overridden by cognitive action. For example, in Anglo-European countries, having offspring is delayed well past the hormonal drive to do so. Condoms, diaphragms, spermicides and birth-control pills have made it possible to bring about that delay. Without the ability to override the hormonal propensity, however, they would not be used.

There are many propensities that we normally do not choose to inhibit (hunger, for example). There are many of which we are not conscious. The propensities that we do choose to inhibit are those that we consider to serve our interests poorly when not mitigated. We can be, and often are, wrong about such things. That does not change the fact that we are cognitive animals and can therefore tame some of our evolved propensities. A sophisticated application of evolutionary theory to human behaviour has to incorporate that fact.

Lewontin, Gould and a number of other professors as well as students formed the Sociobiology Study Group based at Harvard University, and it soon became affiliated with an organization formed in 1969 called Science for the People. A statement on the Science for the People website makes the organization's origin and commitments crystal clear:

> Science for the People arose in 1969 out of the anti-war movement and lasted until 1989. With a Marxist analysis and non-hierarchical governing structure, Science for the People tackled the militarization of scientific research, the corporate control of research agendas, the political implications of sociobiology and other scientific theories, the environmental consequences of energy policy, inequalities in health care, and many other issues.

In light of this, it is obvious that the social and political commitments of Lewontin and Gould aligned well with those of Science for the People.

The political critiques of sociobiology by numerous academics and left-leaning groups were vicious and at times deeply personal. To his credit, Wilson remained civil throughout. It is not surprising that a work as broad-ranging as *Sociobiology: The New Synthesis*, and one, moreover, that was the first attempt at a comprehensive codification of a field of enquiry, would contain overstatements and contestable claims. Controversy was almost inevitable. The malicious nature of the critiques was not.

By the late 1980s the controversy had become muted. More nuanced approaches to evolutionary explanations of behaviour were being given, and many of the criticisms of Wilson's exposition of sociobiology had been accepted and the explanations modified. Today the study of the evolution of social behaviour is pervasive. Courses and research programmes on it exist in departments of biology, psychology, sociology, anthropology and the like. Explaining social behaviours and non-social behaviour using evolutionary theory is widespread. Evolutionary theory has embraced yet another element of the biological world. The synthesis has expanded.

TEN

Development: From Genes to Organisms

We are coming close to the present. For about 100 years the focus of evolutionary theory was on genetics. From time to time there were references to the importance of the process from genes to the adult organism – the process of development – but there was no really sustained emphasis. That changed in the 1970s.

Contemporary attention to the importance of development can reasonably be seen as beginning with Stephen J. Gould's *Ontogeny and Phylogeny* of 1977. In it Gould examined heterochrony and recapitulation, redefining and refining both. Recapitulation relates ontogenetic (embryological) development to phylogenetic (evolutionary) development. Ernst Haeckel is credited with introducing the idea that during ontogenesis the developing organism repeats in fast motion all the stages of its species' evolutionary development. His famous expression is that 'ontogeny recapitulates phylogeny'. This tight connection between ontogeny and phylogeny had become untenable by the early twentieth century, but there are connections between them, and that is part of what Gould explored.

The other idea that Gould explored and redefined is heterochrony: the timing of events during development. For Haeckel, heterochrony was a false pattern of development – one in which ontogeny did not recapitulate phylogeny. Gould used it to describe the timing of changes in shape and size during development. This became the basis for his formulation of 'a compellingly simple clock model for heterochrony'. Biologists have almost universally adopted Gould's definition. In the almost 40 years since Gould's book appeared, it has become clear that the timing of events during embryological development are important to the formation of the adult organism. It also became clear that phylogeny (evolutionary development) and ontogeny were inextricably connected. Evolution and development were now linked. They are linked not in the simplistic way Haeckel postulated but in a complex way that makes changes in developmental timing increase variability. Genetic mutations are one source of new variants but so are 'mutations' in the timing of the development of shape, size and other major features.

This emerging view that development was important to evolution was bolstered by the discovery in 1984 of complexes of genes that control development and switch some genes on and others off. A control process exists. Different species share a significant number of genes. What kind of organism (dog, monkey, human and so on) will emerge depends crucially on which of those shared genes gets used during embryological development. Many of the genes will be unused; which of them are used during development is determined by a set of other genes.

To understand this, let us step back in history. Karl Ernst von Baer and Christian Pander in the latter part of the 1800s discovered that all the organs in a chicken developed from three layers of tissue in the embryo. In fact, we now know that the organs of every animal develop from one of these three layers (ectoderm: outer, mesoderm: middle, and endoderm: inner). Next, Hilde Mangold (a graduate student in Hans Spemann's lab) in the 1920s discovered that a small area of tissue seemed to organize development. The Nobel Prize in Physiology and Medicine for 1935 was awarded to Spemann for this organizing tissue. Mangold died tragically before it was awarded. The notion of an organizer area of tissue during development was exciting but its action remained somewhat elusive.

There were some advances after Mangold and Spemann identified an organizing area of tissue, but the next major breakthrough occurred in 1984. Short sequences of nucleotides (about 180 base pairs long) were discovered to be identical (or virtually so) in all species. These were first discovered by Ernst Hafen, Michael Levine and William McGinnis, working in the lab of Walter Jakob Gehring at the University of Basel, Switzerland. Their discovery was published in *Nature* in 1984 with the title, 'A Conserved DNA Sequence in

Homoeotic Genes of the *Drosophila* Antennapedia and Bithorax Complexes'. The paper opens with:

> Many of the homoeotic genes of *Drosophila* [a fruit fly] seem to be involved in the specification of developmental pathways for the body segments of the fly, so that each segment acquires a unique identity. A mutation in such a homoeotic gene often results in a replacement of one body segment (or part of a segment) by another segment that is normally elsewhere. Many of these homoeotic loci reside in two gene complexes, the bithorax complex and the Antennapedia complex, both located on the right arm of chomosome 3 (3R).

Two months later, in July 1984, the similar discovery of Matthew P. Scott and Amy J. Weiner appeared in the *Proceeding of the National Academy of Sciences (U.S.)*: 'Structural Relationships among Genes that Control Development: Sequence Homology between the Antennapedia, Ultrabithorax, and Fushi Tarazu Loci of Drosophila'.

The regions are considered boxed regions of DNA, and the genes, as the titles of the initial articles suggest, are homoeotic (or the variant spelling 'homeotic', both being derived from the Greek *homoioun* = to make the same). The combination of these two terms gave rise to the designation 'homeobox genes', now commonly shortened to 'Hox genes'.

Building on this work, research revealed that these gene sequences control different aspects of development – the development of limbs, for example. There are eight Hox genes in *Drosophila* and much of the early knowledge about Hox genes came from research on *Drosophila*. Different organisms have different numbers of Hox genes. What is now known is that they control a substantial amount of development (in effect, regulating development). Mangold's organizer had been found and is now well understood.

Organisms that have bodies have many features in common. They typically have a front and back, a top and bottom and a left and right. Most also have heads of some kind at the top end and a neural connection from top to bottom – usually a spinal cord running the length of the back. Also, most have a mouth (or some ingestion orifice) and an anus (or some elimination opening) with a digestive system connecting them. There are also some internal organs for other physiological functions. All this arises during development and is under the control of Hox genes (illus. 43, 44).

The discovery of Hox genes links evolution, genetics and development tightly together, making development an integral part of

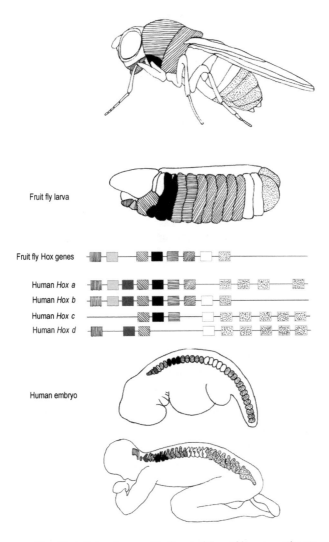

43 Head to tail development in *Drosophila* and humans. The Hox genes shown control the assembling of the segment in the right order.

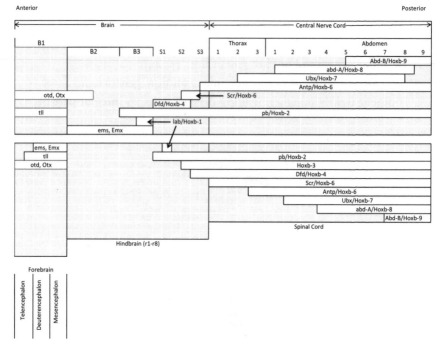

44 A comparison of Hox gene control of brain and neural cord development in vertebrates (bottom frame) and *Drosophila* (fruit flies, upper frame). The focus here is on the Hox gene control of the order of assembly.

the modern synthesis. One movement in the last twenty years that has emphasized and explored the importance of development to evolution has been dubbed evo-devo (evolution plus development). This may have served to bring greater prominence to the role of development in evolution. In light of the role of Hox genes, the terminology seems inappropriate. A better term might be geno-devo, highlighting that the evolution of organisms has shaped both development and genes and both have shaped the evolutionary process. Hence it is not evolution and development but genetics and development in the evolutionary process. Jane Maienschein and Manfred Laubichler capture this well:

> We have argued here that in the context of developmental evolution the causal structure of evolutionary explanation has shifted from a primacy of population-level dynamics to the

primacy of developmental mechanisms and that explaining the origin of variation rather than the fate of variants within populations is the first and most important problem for all theories of phenotypic evolution. Not that evolutionary dynamics are unimportant, rather they act as a (relatively well-understood) filter for variation and as such account for changes over time. But, as many critics of Darwin have pointed out over the last 150 years, selection by itself does not generate variation, and the assumption that mutations, at whatever constant rate, are all that is needed for natural selection to work is no longer a valid argument in light of what we have learned about the complex links between genotype and phenotype. Rather, [we need] an understanding of phenotypic evolution that is grounded in the mechanistic interactions of complex regulatory networks, from genomes to the environment.

Evolutionary theory at the beginning of the twenty-first century encompasses an impressive array of biological domains. Darwin's quest for a consilience of inductions, through which his theory would explain all manner of biological phenomena, has mostly been achieved. One interesting consequence of bringing development fully into the modern synthesis is the re-emergence of the idea that the environment can produce a heritable change in an organism; not in the way Lamarck understood it – that is, in terms of an impact on the adult – but in terms of environmental impacts on developing organisms, impacts that can affect regulatory genes.

ELEVEN

Science and the Death Throes of Biblical Literalism

R obert Ardrey opens his *African Genesis* (1961) with the sentence: 'Not in Innocence, and not in Asia, was man born.' Thus, in ten words, he captures the essence of the chasm between evolution and the Genesis account of creation in the Jewish and Christian scriptures. As we have seen, over the last 150 years the fact of evolution and the theory, descended from Darwin, that explains it have reached the point where doubt seems irrational. There will certainly be further refinements to Darwin's theory but the fact of evolution is indubitable. Nonetheless, as Richard Dawkins has noted, a 2008 Gallup Poll reported that 'More than 40 per cent of Americans deny that humans evolved from other animals, and think that we – and by implication all life – were created by God within the last 10,000 years.' Obviously, the solidifying of Darwin's theory and the mounting evidence for the fact of evolution and a very old earth remains unconvincing to many Americans. Religious fundamentalists, who abound in the United States, reject evolution as incompatible with a literal reading of the Bible. More accurately, it is seen as incompatible with this or that group's literal reading of the Bible since there exists an enormous variety of 'literal readings' within fundamentalist Christianity.

What has always been at stake within Christianity is the place of humans in the cosmos. The Christian Bible is seen as giving pride of place to humans. It also offers the potential for a special relationship with God. Science has inexorably eroded confidence in this view. As Sigmund Freud eloquently put it in his *Introductory Lectures on Psycho-Analysis* of 1917:

In the course of centuries the *naïve* self-love of men has had to submit to two major blows at the hands of science. The first was when they learned that the earth was not the centre of the universe but only a tiny fragment of the cosmic system of scarcely imaginable vastness. This is associated with Copernicus . . . The second blow fell when biological science destroyed man's supposedly privileged place in creation and proved his descent from the animal kingdom and his ineradicable animal nature . . . But human megalomania will have suffered its third and most wounding blow from the psychological research of the present time which seeks to prove to the ego that it is not even master in its own house, but must content itself with scanty information of what is going on unconsciously in its mind.

This captures well the dethroning, by scientific advances, of the special place of humans in the cosmos. We are descended from other animals, live on a planet that is a mere speck in the universe and far from its centre, and have limited free will, if any at all.

There have been two main religious responses to the scientific advances. By far the most dominant is a compatibility stance. The vast majority of Christians (Roman Catholics, Anglicans, Presbyterians, Lutherans and the like) are not biblical literalists and have come to terms with Copernicus/Galileo, evolutionary biology and other scientific advances. The development of this compatability stance has been relatively recent. Historically, the major Christian denominations have tried to thwart scientific advances and ideas. The theories of Copernicus/Galileo and Darwin are prominent and well-known examples.

Just how recent the coalescing of this stance is can be seen in the Catholic Church's handling of the Galileo controversy. It was only in October 1992 that the Galileo Study Commission, which the Pontifical Academy of Sciences established in 1981 at the request of Pope John Paul II, delivered its findings to him. John Paul II gave his reflections on the report in an address to the Plenary Session of the Pontifical Academy of Sciences, and rescinded the Church's edict of Inquisition against Galileo. This is 359 years after it was issued and

449 years after the publication of Copernicus's *De revolutionibus orbium coelestium*. Evolution has fared somewhat better. In the 157 years since the publication of Darwin's *On the Origin of Species*, most major Christian denominations have accepted evolution and found ways to reconcile it with their faith and doctrines, although the success of the various reconciliations is still debated.

Pope John Paul II articulated the Catholic version of the compatibility stance succinctly:

> Both religion and science must preserve their own autonomy and their distinctiveness. Religion is not founded on science nor is science an extension of religion. Each should possess its own principles, its pattern of procedures, its diversities of interpretation and its own conclusions . . . While each can and should support the other as distinct dimensions of a common human culture, neither ought to assume that it forms a necessary premise for the other . . . For the truth of the matter is that the church and the scientific community will inevitably interact; their options do not include isolation. Christians will inevitably assimilate the prevailing ideas about the world, and today these are deeply shaped by science. The only question is whether they will do this critically or unreflectively, with depth and nuance or with a shallowness that debases the Gospel and leaves us ashamed before history. Scientists, like all human beings, will make decisions upon what ultimately gives meaning and value to their lives and to their work. This they will do well or poorly, with the reflective depth that theological wisdom can help them attain or with an unconsidered absolutizing of their results beyond their reasonable and proper limits.

It was his view that theology and science investigate different aspects of our existence and that theology should never again allow itself to be tied to any particular scientific theory or philosophical system. These systems change over time as revolutions in scientific theories and ideas occur. If theology is wedded to a particular theory, should that theory be replaced be by another (Aristotle's by Galileo's, and

Newton's by Einstein's and quantum mechanics, for example), theology's credibility and utility are damaged.

Different major denominations have different versions of this compatibility stance. Nonetheless, today, in some form or another, they all see science and religion as addressing different questions using different methodologies. The major religions in European and English-speaking countries have finally done what dialecticians have always done well: they have found a way through the middle of the horns of a dilemma.

Not so with fundamentalist Christian denominations, whose view of the Bible is literalist. They sit impaled on one horn of the dilemma and refuse to acknowledge the other. What has allowed the major denominations in European countries to adopt a compatibility stance is the slow erosion of a literalist interpretation of the Bible. Although the focus here is mostly on the United States, there is also a regrettable consequence of the colonial period in sub-Saharan Africa. The colonizing missionaries inculcated into the indigenous peoples a literalist interpretation of the Bible – a view that persists today in most parts of sub-Saharan Africa.

There is a plethora of fundamentalist biblical-literalist denominations in the United States, sometimes consisting of only one congregation. They reject evolution and hold to a literal interpretation of the Genesis creation story. As noted, these 'literal' interpretations – for there are many – vary greatly and are inconsistent with each other. This is hardly surprising given that there are two different – and difficult to reconcile – accounts in Genesis of the creation of life and especially humans. The first account is in Genesis 1:24–31, the second in Genesis 2:7–24. Non-literalists simply accept that there are indeed different accounts that are derived from different sources. Literalists do not have this option.

Although the highest-profile hostility of biblical literalists is to biological evolution, their position actually entails the rejection of much of modern science. As Clarence Darrow cleverly pointed out in the Scopes Trial in Tennessee (1925), Joshua could only have lengthened the day by commanding the sun to stand still if the sun revolved around the earth. The heliocentric conception accepted in astronomy at the time of the trial and today would require the earth to stand still

– cease its rotation – for the day to be lengthened. This contradicts a literal reading of Joshua 10:1–15. Moreover, a literal reading of the genealogies, from Adam and Eve up to the present day, requires a very short period since the creation of the earth. Based on biblical genealogy and some other historical material, Bishop James Ussher (1581–1656) dated creation at 4004 BCE (a little more than 6,000 years ago), and the Jewish calendar dates the creation of Adam to 5,775 years ago (as of January 2015). This Jewish dating system is more than 1,600 years old (probably devised by Hillel b. R. Yehuda around 358 CE). Hence no assumptions about modern Judaism's positions on evolution or a very old earth and even older cosmos can be drawn from them. But it does reinforce Ussher's genealogical dating of creation. It appears then that biblical literalists, based on this genealogical calculation, must also reject modern geology and the physics underlying carbon-14 and potassium-argon dating, which show the earth to be about 5.54 billion years old. In one fell swoop biblical literalism relegates contemporary astrophysics, physics, biology and geology – in short, contemporary science – to the dustbin.

This is why, despite all the *political* cacophony and machinations they generate, biblical literalists remain on the *intellectual* fringes. Biological evolution is widely accepted and rightly so. The evidence for it and its conceptual foundations are solid, as our journey has demonstrated. If at some future time a different theory seems more consistent with the evidence (as happened with Newtonian mechanics at the beginning of the twentieth century), then the new theory will be adopted. Today, evolutionary theory, along with quantum theory and general and special relativity, are the theories most consistent with the evidence and provide robust explanations and predictions of the nature and behaviour of things in the world as we experience them. Whatever theories might replace them in the future, they will be responses to the nature and behaviour of things in the world as we experience them, not on biblical literalism.

Given this, the Christian biblical-literalist phenomenon is a mere curiosity in the context of our story. Its failures do not directly strengthen the fact and theory of evolution. The responses of evolutionary biologists to their challenges have clarified elements of the theory and thereby strengthened it. The challenges have mostly focused on passing legislation that either prohibits the teaching of evolution in publicly funded schools – not a successful strategy today – or teaching a so-called scientific alternative to evolution. What follows are a few examples.

The best-known early battle between evolution and biblical literalism was the Scopes trial. Scopes was a teacher who with the support of the Civil Liberties Union challenged a Tennessee law prohibiting the teaching of evolution. Not surprisingly he was found guilty and fined; the fine was later rescinded because the judge erred in aspects of assigning it. It was the dramatic nature of the trial, not the verdict, that was fascinating. As already described, when William Bryans Jennings was on the stand for the prosecution and Clarence Darrow, the defence lawyer, was cross-examining him, Darrow elicited from Jennings the statement that he believed every word in the Bible to be scientifically true. Darrow, in short order, exposed the risible character of this view by asking whether when Joshua commanded the sun to stand still, it was the sun or the earth that actually stopped moving. If the former, science from Copernicus to the time of the trial must be completely wrong: a view that very few would find credible. If the latter, the Bible contains a scientifically inaccurate account of the event. Major East Coast newspapers covered the trial, and Tennessee legislators and their law became a laughing stock as a result.

Arguably the most important contemporary battle centred on an Arkansas law requiring 'balanced treatment'. That is, the 'creation theory' of the origins of life and its variety of forms had to be given equal treatment in the science curriculum with 'evolution theory'. The court struck down the law. The proceedings of the case are as important as the outcome. Both the outcome and the proceedings unequivocally supported evolution and the theory explaining it, and exposed 'creation science' as nothing more than a religious creed, not science at all.

A more recent attempt to reinvigorate the biblical literalist view centres on the complexity of the universe, life and the processes of life; the name given to it is intelligent design. This avoids using the terms 'God' and 'creator'. There are many complex systems in the physical and biological world. In the biological realm, advocates of intelligent design think they see cases where evolution could not be the cause. The examples provided focus on elements of an organism that cannot be assembled in stages, as evolution would require. There has been no shortage of such claims in the past. Mimicry in butterflies and the eye are two well-known ones. In both cases, evolutionary theory can be shown to explain how it occurred and empirical evidence supports that it did evolve. Consider mimicry, where one butterfly mimics the wing patterning of another. The purpose of the mimicry can be protection from predators. The butterfly being mimicked is poisonous or acrid, so many predators (birds, toads, lizards and dragonflies, for example) avoid it. The butterfly that mimics it is also avoided. The claim is that having only part of the pattern will be inadequate; you must have the entire pattern to be mistaken for the one to be avoided. Population dynamics suggest that this is false. Although the easy explanation is a little anthropomorphic, it can be expressed in purely mechanical terms. Here is the easy explanation. When faced with choices, a predator will take a safe bet over a questionable one; a butterfly with even a small part of the pattern of the butterfly to be avoided will be a less safe bet than another kind. The obvious next question is: 'then why did the entire pattern evolve, if a part is enough?' The more a butterfly that is not poisonous resembles one that is, the greater is its protection, especially when food resources for the predator are scarce. If food resources become scarce, a predator might take its chances rather than starve, but it will not be able to differentiate the types. When there is an abundant food supply, even a partial resemblance is good enough for protection. But food supplies wax and wane, so selective pressure decreases and increases. Butterflies that resemble poisonous ones slightly more will do better in environments where food is scarce. Hence the pattern will evolve under mutation and selective pressure.

Ultimately, intelligent design is a 'designer-of-the-gaps' view. Many millennia ago, there were many gaps in our understanding of the world around us and a variety of religious views provided explanations, many involving a deity. The Judaeo-Christian God's creative and intervening power explained many mysteries in parts of the world. Over the last 1,200 years mechanistic explanations for a variety of things have emerged; they are completely physically based, with no appeal to the supernatural. Today, through science, we have explained mechanistically an amazing array of phenomena. The space for God as an explanation has shrunk dramatically. We now have a God-of-the-gaps. God is a substitute for not currently knowing the explanation. Over time the gaps get fewer and the need for God as an explanation shrinks. There are still many challenges but, on inductive grounds, looking to science for the explanation is a better bet than seeking a supernatural one. The problem with intelligent design is that it does not really explain anything. When we face an explanatory challenge with respect to a complex system and we are told it was intelligently designed, we have not really added anything to our understanding. We do not know how it was designed, or who or what designed it, or even why it was designed that way. At least with God, there are some attributes and some purposes but an intelligent designer is just, well, an intelligent designer. One suspects that if the features of an intelligent designer were given, the similarity to the Christian God would be striking. Phillip Kitcher in his book *Coming to Terms with Darwin*, as well as others, has provided in-depth analyses of this view, which undermines its credibility.

The foundation for a non-literalist understanding of the Bible began during the latter part of the nineteenth-century. The shift began, mostly, in Germany with 'higher criticism'. The word 'criticism' is closer to 'analysis' and does not entail a rejection of faith, but it does entail a recognition that a defensible reading of the Bible has to be based on seeing it as text open to historical investigation and contextual interpretation. The books of the Bible are products of their time and were written for specific purposes and for specific audiences. With this approach, one can understand why the Gospels differ on substantial issues.

All this leads to the central problem with biblical literalism. A literalist stance requires that the text speaks plainly and unequivocally to each reader. Any recourse to accounting for, adjusting or explaining apparent inconsistencies opens the landscape for differing 'literal' readings – surely an oxymoron. This is apparent in literalist fundamentalist denominations in the United States today. Since the Bible does not speak plainly and unequivocally to each reader, biblical literalism is completely untenable. This why it began to wane in the latter half of the nineteenth century in England and Europe. Today almost all major religions in European countries have abandoned literalism. Even in the u.s., the major religious denominations have found ways to make compatible faith and evolution. It is the literalist, fundamentalist groups – mostly in the u.s – that have not. One can take some comfort in the fact that, at every turn, science, including evolutionary science, is advancing. As a result, biblical literalists will become more and more marginalized.

APPENDICES

Appendix A: Analysing Data

In large populations, quantitative traits, with a few explainable exceptions, conform to a normal distribution; quantitative traits are those that can be quantified – measured in discrete units – such as height, weight, hairs on a body and cortex mass. The Central Limit Theorem provides the mathematical basis for this; it explains why, when a random variable is the sum or mean of many independent identically distributed random variables, distributions tend to be close to the normal distribution.

As can be seen, the frequency is smooth and is symmetrical above and below the mean (commonly called 'the average' and symbolized by $\bar{x}$).

An important statistical measure is standard deviation. It indicates the amount that the population deviates from the mean. A classic case, scores in an exam, will illustrate this measure. Suppose 50 students write an exam; some will do poorly, some will do very well and the majority will do okay (average). The actual scores are shown in the table overleaf (illus. 45).

The mean score is 5.62. To calculate the variance of the scores from the mean, step one is to subtract each score from the mean (or the mean from the scores). Since those numbers would sum to 0, the next step is to square the result. Then sum these values and divide by the number of students (50 in this case). This yields the standard variance of 4.9556. The standard deviation is the square root of the variance ($\sqrt{4.9556} = 2.226$). Hence, 1 standard deviation above the mean is 7.846, and 1 standard deviation below the mean is 3.394.

Fifty is a small group. If there were 500,000 students, the scores, consistent with the Central Limit Theory, would conform to a normal curve. If the lowest score were 2/10 and the highest 9/10, the mean would be 5.5 and the standard deviation would be 1.75, which means 68.27 per cent of the students would have scores between 3.75 and 7.25. Different distributions of scores, traits and so on will have different mean values, but if they conform to a normal distribution, 68.27 per cent of the population will be within one standard deviation above or below the mean.

Person 35	4	−1.62	2.6244
Person 38	4	−1.62	2.6244
Person 39	4	−1.62	2.6244
Person 45	4	−1.62	2.6244
Person 2	5	−0.62	0.3844
Person 5	5	−0.62	0.3844
Person 9	5	−0.62	0.3844
Person 17	5	−0.62	0.3844
Person 31	5	−0.62	0.3844
Person 33	5	−0.62	0.3844
Person 41	5	−0.62	0.3844
Person 44	5	−0.62	0.3844
Person 47	5	−0.62	0.3844
Person 50	5	−0.62	0.3844
Person 7	6	0.38	0.1444
Person 15	6	0.38	0.1444
Person 20	6	0.38	0.1444
Person 27	6	0.38	0.1444
Person 32	6	0.38	0.1444
Person 40	6	0.38	0.1444
Person 43	6	0.38	0.1444
Person 6	7	1.38	1.9044
Person 19	7	1.38	1.9044
Person 26	7	1.38	1.9044
Person 34	7	1.38	1.9044
Person 42	7	1.38	1.9044
Person 46	7	1.38	1.9044
Person 8	8	2.38	5.6644
Person 14	8	2.38	5.6644
Person 21	8	2.38	5.6644
Person 36	8	2.38	5.6644
Person 37	8	2.38	5.6644
Person 3	9	3.38	11.4244
Person 12	9	v3.38	11.4244
Person 13	9	3.38	11.4244
Person 23	9	3.38	11.4244
Person 29	9	3.38	11.4244
Person 30	9	3.38	11.4244
Person 48	9	3.38	11.4244
	5.62		4.9556

45 A table of scores on a test arranged by score (ascending). The final column gives the deviation of that score from the mean.

Appendix B: Hardy-Weinberg Proof

The proof of the Hardy-Weinberg equilibrium is fairly simple. Assume two factors A and B; also assume, in the initial generation (F_0) p = the proportion of A factors and q = the proportion of B factors. Construct a breeding matrix (assuming random mating) (illus. 46).

Focus on the second column and second row (p^2AA). The offspring from the union (fertilization) of an ovum with the factor A and a sperm also containing factor A will be AA. The proportion of A in the population is p. Hence when the ovum and sperm combine, the proportion is $p \times p$, which equals p^2. The same is true for the third column and third row, except the factor is B and the proportion q. The same multiplication process yields the second column, third row, and third column, second row.

AB is the same as BA in terms of its genetic effect, so there will be $2 \times pq$ of this combination. Hence, the ratios after mating (i.e., in the first generation, F_1) are: $p^2AA : 2pqAB : q^2BB$. To move to the next generation, we need to determine the ration of A to B in this first generation. With respect to A, there are $2p^2$ of them (because there are 2 As with a proportion of p^2) + $2pq$. With respect to B, there are $2pq + q^2$.

Hence:

$$A{:}B = 2p^2 + 2pq{:}2q^2 + 2pq$$

$$2p^2 + 2pq = 2p(p + q), \text{ and}$$

$$2q^2 + 2pq = 2q(p + q)$$

Hence:

$$A{:}B = 2p(p + q){:}2q(p + q)$$

Dividing (or multiplying) both sides of a ratio by the same thing does not change the ratio. So, now divide both sides by $(p + q)$:

$$2p{:}2q$$

Now, divide both sides by 2:

$$p{:}q$$

	p (A)	q (B)
p (A)	p² (AA)	pq (AB)
q (B)	pq (AB)	p² (BB)

46 The mating of hybrids p and q are the proportion of an allele (Mendel's factor) in the breeding population.

Which was the initial ratio of *p:q*. Since that ratio yielded $p^2AA : 2pqAB : q^2BB$ in the first generation, it will do so again in the second generation. Moreover, the calculation of the ratio in the third generation will yield the same as in the second generation, and so on. The ratio will remain constant forever. Contrary to Punnett, dominance has no effect on the ratios in subsequent generations and, more importantly, there is no specific ratio to which a randomly mating population will converge. Punnett's claim that one would expect 3:1 is wrong; all that matters is the starting ratios, after which every generation will be the same.

Appendix C: R. A. Fisher's Death Rate and Reproduction Rate Equations

For death rate, if l_x = the number in a population living to age x, the death rate at age x (chance of someone dying at age x) is:

$\mu_x = - (1/l_x \, dl/dx) \, l_x$, which equals, using Fisher's notation: $-d/dx \, (\log l_x)$

For the reproduction rate, if b_x = the rate of reproduction at age x, the chance of someone living to reproduce during the interval dx is:

$$\rho_x = l_x b_x dx$$

This can describe the total expectation of reproduction for an organism at birth by integrating over the lifespan of the individual from birth (0) to death (∞); this captures every age during the organism's lifetime in which it could reproduce:

$$\int_0^\infty l_x b_x dx$$

The relative rate of increase can be included by adding a 'Malthusian' variable m. Then, as Fisher points out, 'the number of persons in the infinitesimal age interval dx will be proportional to $e^{mx} l_x b_x dx$, for of those born only a fraction l_x survive to this age'. The aggregate for all ages is:

$$\int_0^\infty e^{-mx} l_x b_x dx$$

To this one can add a valuation that recognizes that, except in a steady state, every age in the life cycle will have a different valuation of reproduction, v_x. Fisher's all in equation for reproductive value is:

$$n_x \{(b_x v_0 - \mu_x v_x) \, dx + dv_x\}$$

From this, by differentiation, one gets:

$$dv_x - \mu_x v_x dx + b_x v_0 dx = m v_x dx$$

Appendix D: Fisher's Mathematical Expression of Variation and its Relation to Fitness

The two groups of individuals bearing alternative genes, and consequently the genes themselves, will necessarily either have equal or unequal rates of increase, and the difference between the appropriate values of m will be represented by a; similarly the average effect upon m of the gene substitution will be represented by a. Since m measures fitness to survive by the objective fact of representation in future generations, the quantity $pqaa$ will represent the contribution of each factor to the genetic variance in fitness; the total genetic variance in fitness being the sum of these contributions, which is necessarily positive, or, in the limiting case, zero. Moreover, any increase dp in the proportion of one type of gene at the expense of the other will be accompanied by an increase $a\,dp$ in the average fitness of the species, where a may of course be negative; but the definition of a requires that the ratio p:q must be increasing in geometrical progression at a rate measured by a, or in mathematical notation that

$$d/dt \log (p/q) = a$$

which may be written:

$$(1/p + 1/q)dp = a\,dt,$$

or
$$dp = pqa\,dt$$

Whence it follows that,

$$a\,dp = pqaa\,dt$$

And, taking all factors into consideration, the total increase in fitness is:

$$\Sigma\,(a\,dp) = \Sigma(pqaa)dt = W\,dt$$

When dt is positive, $W\,dt$ will also be positive. In his 1958 revised version, Fisher provides additional steps to the final equation: he uses α rather than the smaller a, which avoids confusion of the two as (a and a) in the 1930 exposition, and, since $p + q = 1$, he expresses equation in terms solely of p: $\Sigma\,(a\,dp) = dt\Sigma\Sigma'(2paa) = W\,dt$.

Fisher then asserts that: 'the rate of increase in fitness due to all changes in gene ratio is exactly equal to the genetic variance of fitness W which the population exhibits'.

Appendix E: Fisher's Incorporation of Fitness

For this exposition, I use the current convention of w for fitness and $\bar{w}$ for average fitness. If a_1a_1 has a fitness of w_{11}, a_1a_2 has a fitness of w_{12}, and a_2a_2 has a fitness of w_{22}, then, after selection:

$$w_{11}(p2)a_1a_1 : w_{12}(2pq)a_1a_2 : w_{22}(q^2)a_2a_2$$

Normalizing to make $p + q = 1$, yields:

$$(w_{11}(p^2)/\bar{w})a_1a_1 : (w_{12}(2pq)/\bar{w})a_1a_2 : (w_{22}(q^2)/\bar{w})a_2a_2$$

$$\text{where } \bar{w} = \text{average fitness} = w_{11}(p^2) + w_{12}(2pq) + w_{22}(q^2)$$

TIMELINE

550 BCE Anaximander of Miletus writes *On Nature*, in which he proposes the evolution of organisms and that life began in oceans

1616 Lucilio Vanini is burned alive at the stake in France for proposing that humans evolved from apes

1658 Bishop J. Ussher's *The Annals of the World* is published. In it, based on ancient histories and the Bible, he dates the first day of creation as Sunday, 23 October, 4004 BCE (although he used the older designation BC), the expulsion of Adam and Eve from the Garden of Eden as Monday, 10 November 4004 BCE, and Noah's Ark landing on Mt Ararat as 5 May 2348 BCE

1794 Volume I of Erasmus Darwin's massive two-volume *Zoonomia; or, The Laws of Organic Life*, in which he takes an evolutionary perspective, is published

1796 Volume II of Erasmus Darwin's *Zoonomia* is published

1799 Friedrich Schleiermacher's *On Religion: Speeches to its Cultured Despisers* is published. This is the beginning of biblical higher criticism (Tübingen, Germany)

1801 Jean-Baptiste Lamarck tentatively proposes a view of evolution in *Système des animaux sans vertèbres*

1809 Jean-Baptiste Lamarck, in *Philosophie Zoologique*, gives his more mature views on the evolution of animals from simpler to more complex

1835   *The Life of Jesus Critically Examined* by David Friedrich Strauss is published, part of the movement to uncover the historical Jesus. Strauss challenges biblical literalism, denies the divinity of Jesus and claims that myth and legend so obscure the Gospel accounts of Jesus that constructing a historical picture of Jesus is impossible

1837   On 1 July Charles Darwin begins his evolutionary notebook 'on the variation of animals and plants under domestication and nature'

1839   Charles Darwin's *Voyage of the Beagle* is published

1842   Darwin writes a 35-page abstract of his theory

1844   Robert Chambers anonymously publishes *Vestiges of the History of Creation*; a book filled with evolutionary speculations

Darwin expands his abstract to 230 pages

1844–58 Darwin expands the manuscript considerably and, by 1858, he has a very large manuscript, which is now known as his 'big book on species'

1846   George Eliot's English translation of Strauss's *Life of Jesus* is published

1853   George Eliot's English translation of Feuerbach's *Essence of Christianity* is published

1855   Alfred Russel Wallace's paper 'On the Law which has Regulated the Introduction of New Species' is published in *Annals of Natural History*

1857   On 10 October Darwin receives a letter and paper from Alfred Russel Wallace, from the island of Celebes in the Malay Archipelago (today Sulawesi, Indonesia). In the paper, written, according to Wallace, on Ternate, Indonesia, he sets out a version of natural selection very similar to Darwin's

1858   The papers of Darwin and Wallace on natural selection are read to the Linnaean Society

1859   Charles Darwin's *On the Origin of Species* is published

1865   Gregor Mendel publishes 'Experiments on Plant Hybrids'

1890   W.F.R. Weldon publishes his first paper on variation, marking the beginning of the movement to treat variation statistically

1894   William Bateson publishes *Materials for the Study of Variation*. It expounds a saltation view of evolution; evolution occurs through major transformational leaps, not, as Darwin claimed, gradually

Karl Pearson publishes 'Contributions to the Mathematical Theory of Evolution', which builds on the views of W.F.R. Weldon, with whom Pearson by this time has become good friends. This view becomes known as the biometric view of heredity and is opposed to the Mendelian view

1900   Hugo de Vries and Karl Correns produce Mendel's results, 'rediscovering' Mendel's First Law (both acknowledge Mendel's priority)

William Bateson coins the term 'genetics'

1902   Walter Sutton, a graduate student at Columbia University, in a single offhand sentence, connects the observed behaviour of chromosomes with Mendel's mathematical account of his hereditary factors

1905   William Bateson shows that traits are not always inherited separately but are linked – always occurring together

1908   G. H. Hardy and Wilhelm Weinberg independently publish a proof of the equilibrium principle for Mendelian populations

1924   J.B.S. Haldane publishes part 1 of 'A Mathematical Theory of Natural and Artificial Selection' (between then and 1932 he publishes another nine parts)

1916–17 Sigmund Freud's three-volume *Introductory Lectures on Psycho-analysis: A Course of Twenty-eight Lectures Delivered at the University of Vienna* is published

1925   John T. Scopes, a high school teacher in Tennessee, is charged, tried and convicted of teaching evolution

1926   Based on extensive breeding experiments and microscope-based examinations, Thomas Hunt Morgan publishes *The Theory of the Gene*, which provides a comprehensive account of Mendelian genetics and resolves Bateson's linked trait observations

1930   Ronald A. Fisher publishes *The Genetical Theory of Natural Selection*

1931   Sewall Wright publishes 'Evolution in Mendelian Populations'

1932   Population genetics, based on the work of Haldane, Fisher and Wright, is a robust and increasingly accepted modelling of the genetics of populations. The views of the biometricians and Mendelians are synthesized in population genetics

1936   Andrei N. Belozersky isolates pure deoxyribonucleic acid (DNA)

1937   William Astbury uses X-ray diffraction to study nucleic acid

1944   Oswald Avery (working with Colin MacLeod and Maclyn McCarty) shows that DNA is the hereditary material

        George Gaylord Simpson publishes *Temp and Mode in Evolution*

1946   Joshua Lederberg observes and describes genetic recombination in the bacterium *Escherichia coli* (*E. coli*)

1950   Alec Stokes, using X-ray diffraction images of DNA produced by Maurice Hugh Frederick Wilkins, works out that DNA most likely has a helical structure

        Erwin Chargaff discovers that, in a molecule of DNA, there are an equal number of adenine and thymine bases and an equal number of cytosine and guanine bases

1952   Alfred Hershey and Martha Chase prove that DNA is the genetic material involved in hereditary transmission

1953   James D. Watson and Frances H. C. Crick, using the X-ray diffraction images of Rosalind Elsie Franklin and Maurice H. F. Wilkins, model the structure of DNA as a double helix. This unravels the mystery of genetic coding and the molecular basis of heredity

1959   Mary N. Leakey, during archaeological work in the Oldupai Gorge in Tanzania (frequently called Olduvai due to early mispronunciation of Oldupai) finds the jaw, and shortly after the skull, of *Australopithecus boisei*. The skull was found to be 1.75 million years old

1961   Marshall Nirenberg demonstrates that a triplet of the base uracil (UUU) in RNA specifies (codes for) the amino acid phenylalanine. That is, three uracil bases adjacent on a strand of RNA results in phenylalanine being inserted in a protein under construction

1966   Marshall Nirenberg, building on his early discovery (see 1961), proves that most triplets of bases in a strand of RNA is associated with (codes for) a specific amino acid (the building blocks of proteins), resulting in the creation of a coding dictionary

1972   Stephen Jay Gould and Niles Eldridge publish an article in the volume *Models in Paleobiology* (ed. T.J.M. Schopf) advocating a view of evolutionary change called punctuated equilibria

        Masatoshi Nei develops a method for genetic estimating distances between populations from electrophoretic data

1974   Donald Johanson leads a team of archaeologists and geologists in an exploration of the Hadar Formation in the Afar Triangle of Ethiopia and discoveries the skeleton of *Australopithecus afarensis*, dated as 3.2 million years old. *A. afarensis* had a small, ape-size, cranium but was bipedal

1975   Edward O. Wilson, already famous for the discovery of pheromones in ants, publishes his controversial *Sociobiology: The New Synthesis*, in which he argues that many animal behaviours, including many human behaviours, have a biological evolutionary basis

1977   Stephen J. Gould's book *Ontogeny and Phylogeny* is published

1981   Charles G. Sibley and Jon E. Ahlquist use DNA analysis to develop an evolutionary history of flightless birds

        The Legislature of the State of Arkansas passes a law mandating 'balanced treatment' of evolution and creation science in public schools

1982   Judge William R. Overton, in Little Rock, Arkansas, rules that the
        Balanced Treatment Act is unconstitutional

1983   Walter Jakob Gehring and his research team discover a sequence of
        genes – homeobox genes (Hox genes) – that direct embryological
        development

1985   Charles G. Sibley and Jon E. Ahlquist use DNA analysis to develop an
        evolutionary history of Australian birds – perching and songbirds –
        which results in revisions to the existing history

1997   Ian Wilmut and his research team at the Roslin Institute at the
        University of Edinburgh clone a sheep, the first mammal to be
        cloned, and name it Dolly

1998   S. Kumar and S. B. Hedges find excellent agreement between fossil-
        based estimates of evolutionary divergence times for major vertebrate
        lineages and combined molecular divergence estimates

GLOSSARY

Actualism: The idea that the causes of geological events in the past are the same as the ones observed in action today.

Amino Acids: The building blocks of proteins. There are twenty essential amino acids. The order of the amino acids in a linked sequence determines the nature and action of the protein.

Antibodies: Entities in the immune system that are created in response to an antigen (foreign body: viruses and bacteria, for example). Antibodies engulf the foreign material and destroy it. Once formed, antibodies can last for long periods, sometimes the entire life of the organism. This confers immunity against the specific foreign material.

Antigens: Material foreign to an organism, frequently viruses, bacteria and parasites. These trigger a response from the organism's immune system.

Antiserums: Antibodies that have developed in response to specific antigens.

Axioms: General statements (laws) in theory that cannot be deduced from any other general statements. In a mature theory, all other general statements can be deduced from its axioms.

Binomial Distribution: The distribution that results from repeated occurrences of events with only two (*bi*nomial) outcomes. A coin tossed many times will come up heads or tails each time it is tossed. For a fair coin, the outcome heads will equal the outcome tails over a very large number of tosses. The binomial distribution is the distribution of the probability of a particular outcome: all heads, all tails or (over 30 tosses) 25 heads and 5 tails, for example. If a coin is tossed 2 times, the probability of any particular outcome can be calculated by expanding the binomial $(H + T)^2$. That is, by multiplying $(H + T)$ by itself 2 times = $HH + 2HT + TT$. If a coin is tossed 30 times, the probability of any particular outcome can be calculated by expanding the binomial $(H + T)^{30}$: multiplying $(H + T)$ by itself 30 times. The result can be graphed.

Biometricians: The name of a group of researchers who held that traits of organisms can be quantified (measured) and that these traits conform to a normal curve (bell curve). Height is an example. The heights of individuals range from very short to very tall, with most people being in the middle of the range.

Biometry: A field of research in which traits of living (bio) things are measured (metry). Traits such as height, weight, muscle mass, finger length and so on are measurable and vary greatly in a population. Studying the distribution of these differences is biometry.

Blending Inheritance: The theory that hereditary material (for Mendel, factors; today, alleles and genes) blends together in reproduction. This results in the traits of the parents blending together in the offspring. Compare with particulate inheritance.

Chromosome: A hereditary chemical. Chemical strands are normally found in pairs in the nucleus of a cell. During the creation of new cells, the paired strands separate, then duplicate, resulting in each of the original paired strands now having a twin. The cell then pinches in the centre and divides into two cells. One of each twin goes into a separate cell.

Consilience of Inductions: A concept found in the philosophy of science of William Whewell. The more separate domains of inquiry a theory encompasses (explains), the more probable it is that the theory is true. Darwin's theory explained phenomena in biogeography, anatomy, palaeontology and the like. Newton's theory explained astronomical phenomena, oceanographical phenomena (tides, for example), physical interaction (billiard balls on a billiard table, for example) and so on.

Correlation: The degree to which two (or more) things (traits, events and so on) are related. Two traits or events that always occur together have a correlation of 1. Two traits or events that never occur together have a correlation of 0. Income and health are correlated in the 0.35 to 0.47 range (see S. J. Babones). This is a strong association.

Cytology: The study of the structure and functioning of cells.

Cytogenetic: The genetics of cells.

Deoxyribonucleic Acid (DNA): The chemical of heredity. It contains the genetic code of individual organisms.

Development: *see* embryology.

Dizygotic Twins: Twins that form from two separate zygotes (fertilized eggs)

DNA: See deoxyribonucleic acid.

Embryology (embryological development; development): The process from a fertilized ovum (zygote) to an independently functioning organism. In humans, this is from conception to birth.

Empiricism (empiricist): The view that all knowledge of the world is derived from experience, through the five senses. Our knowledge of cause and effect goes beyond experience, since we do not observe it. We only observe one event being followed by another constantly. We conclude that the first must be causing the second. But that is an inference, not an observation.

Epicycles: In Ptolemy's astronomy, there were cycles above the main cycle of the planets. His astronomy pre-dated Copernicus, and the earth was the centre of the universe. Jupiter circled around the earth; that was its main orbit. But it also circled around a point on that main orbit. Hence there was a circle on a circle, a cycle on a cycle – an epicycle.

Epistemology: The study of what can be known and how it is known (from the Greek, *episteme*: knowledge).

Eugenics: Literally means 'good genes'. It is the selecting of individuals with 'good genes' to breed and the stopping of those with bad genes from breeding. Sterilizing mentally challenged individuals is an example of eugenics.

Eukaryotes: Organisms whose cells have a nucleus. The nucleus encloses the chromosomes – chromosomal DNA. Compare with prokaryotes.

Gametes (gametic cells): Sex cells: sperm/pollen and ova. These cells have one half of the normal number of chromosomes. After fertilization (the union of two gametes) the resulting cell has the normal number of chromosomes.

Genera: *see* genus.

Genetic Determinism: The view that genes determine all aspects of an organism, including its behaviour and, in an extreme form, its beliefs.

Genetic Drift (Random Genetic Drift): Random genetic drift occurs as a result of sampling during mating. More of some genes make it into the mating pool.

Genus (genera): The grouping of similar species; the level of grouping immediately above the species level. The biologist Linnaeus developed a binomial (two-name) system of classifying organisms. The first term is the genus to which the organism belongs. The second term is the species. Humans are *Homo sapiens*, for example: we belong to the genus *Homo* and the species *sapiens*.

Germ Cells: The same as gametes.

Germ Plasm: The hereditary material. The same as gametes.

Glumes: A husk around a seed. Commonly referred to as chaff.

Gradualism: The view that Darwin espoused, which claims that evolution occurs through the gradual accumulation of small changes. Evolution occurs through natural selection acting on small individual variations in organisms within an interbreeding population. This view is opposed to saltationism.

Heredity (heritability): Characteristics of parents are passed to their offspring.

Hermaphrodites: Organisms that are self-fertilizing.

Heterochrony: The timing of events during development.

Homologies: Broad classes of organisms (mammals, for example) that share common features. Anatomically, the hand of a human, the wing of a bat, the toe of a shrew and the like share a similar structure. Some bones are longer in one than another but they are all there.

Homologous Chromosomes: A matched pair of chromosomes that complement each other.

Hypertrophy: From Greek hyper, 'excess', and trophē, 'nourishment'. The enlargement of an organ resulting from increase in the size of its cells.

Independent Assortment: Mendel's principle that the genetic factors (alleles) behave independently of each other when gametes (sex cells) are formed. That is, one allele from a pair will go into one gamete, the other into another, and which goes in which gamete is not affected by other alleles, including its complementary allele.

Isotope: An isotope of an atom is a variant because of a different number of neutrons (particles with no electric charge). Isotopes of an atom will have the same number of protons (positively charge particles) and electrons (negatively charged particles).

Lamarckism: The view that changes in an organism resulting from environmental factors can be inherited. A common example is the giraffe's neck. Giraffes reaching higher to get food from trees developed longer necks. Each generation inherited a long neck.

Linkage: Genes that are close together on a chromosome are linked and are passed to offspring together.

Meiosis: The process by which gametes (sex cells are created). A cell duplicates its chromosomes and then divides into two cells, each with a complete set of chromosomes. Each of those cells divides again, resulting in four cells, with each cell containing half the normal number of chromosomes.

Mendelians: The name of a group of researchers who adopted Mendel's theory of heredity. More specifically, this group rejected Darwin's view that natural selection acts on small individual variations.

Mitosis: The process by which cells divide to create new cells. A cell duplicates its chromosomes and then divides into two cells, each with a complete set of chromosomes.

Monozygotic Twins: Twins that result from a single zygote (fertilized egg). Usually known as identical twins.

Mutability: Literally, the ability to change. In the context of evolution, it is the ability of one species to change sufficiently over time that it becomes a new species.

Natural Selection: The propensity of organisms with a trait to have more offspring than those without the trait. Metaphorically, nature is selecting those with the trait. Today it is common to define 'natural selection' as 'differential reproductive success'.

Newtonian Mechanics (Newton's mechanics; Newton's theory): The physical mechanisms that Newton developed to explain the behaviour of objects.

Nucleotide: The building blocks of DNA and RNA. There are five: adenine, guanine, cytosine, (only in DNA) thymine and (only in RNA) uracil.

Ontogenetic/Ontogeny: The development of an organism from fertilization to birth – sometimes until maturity.

Pangenesis: An early theory of heredity. On this theory, every cell of the body contributes gemmules to the sex cells. These gemmules give rise to the various organs of organism in the next generation.

Particulate Inheritance: This means that hereditary material (for Mendel, factors; today, alleles and genes) remains constant from generation to generation. There is no blending of the hereditary material.

Phenotype: The physiology (and, for some, behaviour) of the adult organism.

Phylogenetic: Evolutionary history of a species. Compare ontogenetic.

Pleiotropic Genes: Genes that produce more than one trait.

Polyploid (Polyploidy): Having three or more matched chomosomes. Compare with diploid.

Population Genetics: A theory describing the behaviour of genes in an interbreeding population.

Prokaryotes: Organisms whose cells do not have a nucleus. The chromosomes (chromosomal DNA) are free within the cell. Compare with eukaryotes.

Protein: The chemical of life. All cells are made of proteins (structural proteins) and all the activities of cells are carried out by proteins (functional proteins). Enzymes are proteins.

Random Sample: A group of things selected in such a way that any group of the same size is equally likely to be selected.

Randomization: Selecting things in such a way that there is no pattern to the selecting. Or, selecting one thing from a collection of things such that each thing in the collection is equally likely to be selected. Commonly understood to eliminate bias.

Randomized Controlled Trials (RCTs): An experimental method in which individuals or other entities are randomly assigned to one of two groups. Those in one of the groups receive an intervention (a pharmaceutical, for example). Those in the other group receive a placebo (a pill that looks like the pharmaceutical but has no active properties). This latter is a 'control' group. If those in the intervention group do better than those in the control group, the intervention is deemed efficacious.

Rationalism: The view that knowledge is a product of reason. Observation is secondary to reason in acquiring knowledge.

Recapitulation: A term used to describe the extent to which the sequence of events in the development of an organism from a fertilized egg to an organism separate from its parent mirrors the evolutionary history of that organism species.

Replication: For experiments: the repeating of an experiment. For organisms: having offspring.

Ribonucleic Acid (RNA): A chemical similar to the strands of DNA, except that it has one different nucleotide: uracil instead of thymine.

Saltationism: The view that evolution occurs through large changes in some organisms in an interbreeding population. Natural selection acts on mutations that are different from the normal organism in a population. This view is the opposite of gradualism.

Somatic Cells: Cells that make up all the organs of a body: compare with gametes/gametic cells. These cells, normally, have chromosomes in pairs.

Special Creation: The view that species of organisms were created, usually by a deity, and did not evolve.

Speciation: The evolutionary process by which new species are created. There are three main kinds: allopatric, sympatric and parapatric. Allopatric is the most common. It occurs when a population is divided by a geographic formation, so the members of the separate groups no longer interbreed and evolve separately. Sympatric speciation occurs when the members of a subpopulation within a larger population interbreed with each other more frequently than with members of the larger population. Parapatric speciation occurs when a geographic barrier is permeable, so there is some interbreeding between geographically separated groups. The groups will evolve in different ways and, at some point, those that breed across the groups will produce less fit offspring.

Species: Organisms that are similar in important respects, which therefore are understood as a group. This is a much-debated concept with a variety of views about how to determine the boundaries of species.

Syllogism: A pattern of reasoning in which there are premises (statements held to be true; usually there are two) and a conclusion (a statement that is claimed to be deducible from – proved true by – the premises).

Theory: A framework that unifies all the knowledge in a domain of science. Newtonian mechanics is an example, as is contemporary evolutionary theory. Theories are used to explain and predict phenomena.

Thermodynamics: The branch of physics that explores heat, temperature and energy of entities and collections of entities.

Transcription: The process whereby messenger RNA is created using the genetic code of DNA.

Translation: The process of creating a protein by linking together a chain of amino acids using the genetic code of messenger RNA.

Transmutation: When one species changes over time to such an extent that it is a new species.

Uniformitarianism: The causes operating in the past were of the same intensity as those observed today.

Valence: A property of an atom that determines its ability to join together with other atoms. The valence of an atom is determined by the number of hydrogen atoms with which it can join. Oxygen has a valence of two because it can join with two hydrogen atoms, as it does in the case of water, H_2O. The number of hydrogen atoms with which an atom can combine is only an index. Atoms can join together in multiple combinations, some including hydrogen, some not.

Variance: The spread of values from the mean (average). For example, a quantity such as height will have an average for a population. There will be people who vary from that average: that is, taller or shorter. Variance is a measure of how height is distributed above and below the mean.

Variety: A way of sub-grouping organisms within a species. There will be a number of varieties of a single species.

Vascular Plants: Plants that have a system of fluid-filled vesicles: similar to blood in animals.

BIBLIOGRAPHY

Ardrey, Robert (1963), *African Genesis: A Personal Investigation into the Animal Origins and Nature of Man*. New York: Dell

Avery, O. T., C. M. MacLeod and M. McCarty (1944), 'Studies on the Chemical Nature of the Substance Inducing Transformation of Pneumococcal Types: Induction of Transformation by a Deoxyribonuleic Acid Fraction Isolated from Pneumococcus Type III', *Journal of Experimental Medicine*, 79(2): 137–58

Babones, S. J. (2008), 'Income Inequality and Population Health: Correlation and Causality', *Social Science and Medicine*, 66: 1614–26

Baer, K. E. Von (1828–37), *Über Entwicklungsgeschichte der Thiere Beobachtung und Reflexion* (2 vols). Königsberg: Bornträger

Bateson, William (1894), *Materials for the Study of Variation Treated with Especial Regard to Discontinuity in the Origin of Species*. London: Macmillan

— (1913), *Problems of Genetics*. Oxford: Oxford University Press

—, and Miss E. R. Saunders (1902), *Reports to the Evolution Committee of the Royal Society*, Report 1. London: Harrison & Sons

Castle, W. E., and Hansford MacCurdy (1907), *Selection and Cross-breeding in Relation to the Inheritance of Coat-pigments and Coat Patterns in Rats and Guinea-Pigs*. Washington, DC: Carnegie Institution of Washington Publications

Chambers, Robert (1845), *Vestiges of the Natural History of Creation*. New York: Wiley & Putnam

Chargaff, Erwin (1950), 'Chemical Specificity of Nucleic Acids and Mechanism of their Enzymatic Degradation', *Experientia*, 6: 201–9

Copernicus [1543] (1976), *On the Revolution of the Heavenly Spheres*, trans. A. M. Duncan. New York: Barnes & Noble

Crow, James F. (2002), 'Perspective: Here's to Fisher, Additive Genetic Variance and the Fundamental Theorem of Natural Selection', *Evolution*, 56(7): 1313–16

Darwin, Charles (1859), *On the Origin of Species by Means of Natural Selection*. London: John Murray

— (1868), *The Variation of Animals and Plants under Domestication*. London: John Murray

— (1871), *The Descent of Man and Selection in Relation to Sex*. London: John Murray

Darwin, Erasmus (1794–6), *Zoonomia; or, The Laws of Organic Life* (2 vols). London: J. Johnson

Darwin, Francis, ed. (1892), *Charles Darwin: His Life Told in an Autobiographical Chapter and in a Selected Series of his Published Letters*. New York: Appleton; reprinted in 1958 under the title *The Autobiography of Charles Darwin and Selected Letters*. New York: Dover

— (1959), *The Life and Letters of Charles Darwin* (2 vols). New York: Basic Books

De Vries, Hugo [1901] (1909), *Mutation Theory: Experiments and Observations on the Origin of Species in the Vegetable Kingdom*, trans. J. B. Farmer and A. D. Darbishire. London: Kegan Paul, Trench, Trünber & Co.

Dobzhansky, Theodosius (1937), *Genetics and the Origin of Species*. New York: Columbia University Press

— (1973), 'Nothing in Biology Makes Sense Except in the Light of Evolution', *American Biology Teacher*, 35(3): 125–9

East, Edward M. (1910), 'A Mendelian Interpretation of Variation that is Apparently Continuous', *American Naturalist*, 44: 65–82

— (1918), 'The Role of Reproduction in Evolution', *American Naturalist*, 52: 273–9

Eldredge, Niles, and S. J. Gould (1972), 'Punctuated Equilibria: An Alternative to Phyletic Gradualism', in *Models in Paleobiology*, ed. T.J.M. Schopf, 82–115. San Francisco: Freeman Cooper

Fisher, Ronald A. (1930), *The Genetical Theory of Natural Selection*. Oxford: Oxford University Press

Freud, Sigmund (1916–17), *Introductory Lectures on Psycho-analysis: A Course of Twenty-eight Lectures Delivered at the University of Vienna*. Authorized English translation by Joan Riviere (1922). London: Allen & Unwin (quotation is from Lecture XVIII, 'Fixation to Traumas: The Unconscious', 1917 [SE, XVI, 284–5])

Galton, Francis (1869), *Hereditary Genius: An Inquiry into its Laws and Consequences*. London: Macmillan

Gayon, Jean (1998), *Darwinism's Struggle for Survival: Heredity and the Hypothesis of Natural Selection*. Cambridge: Cambridge University Press

Gillham, Nicholas W. (2001), *A Life of Sir Francis Galton: From African Exploration to the Birth of Eugenics*. New York: Oxford University Press

Gould, Stephen J. (1977), *Ontogeny and Phylogeny*. Cambridge, MA: Belknap Press of Harvard University Press

Griffith F. (1928), 'The Significance of Pneumococcal Types', *Journal of Hygiene*, 27(2): 113–59

Hafen, Ernst, Michael Levine and Walter J. Gehring (1984), 'Regulation of Antennapedia Transcript Distribution by the Bithorax Complex in Drosophila', *Nature*, 307: 287–9

Haldane, J.B.S. (1924–32), 'A Mathematical Theory of Natural and Artificial Selection' (9 parts), *Transactions* and *Proceedings of the Cambridge Philosophical Society*

— (1932), *The Causes of Evolution*. London: Longmans, Green & Co.

Hamilton, W. D. (1964). 'The Genetical Evolution of Social Behaviour, I', *Journal of Theoretical Biology*, 7: 1–16

— (1964). 'The Genetical Evolution of Social Behaviour, II', *Journal of Theoretical Biology*, 7: 17–52

Hansel, Elise (1908), 'Verebung bei Ungeschlechtlicher Fortpflanzung van Hydra Grisea', *Jenaische Zeitschrift*, 43: 322–72

Hardy, G. H. (1908), 'Mendelian Proportions in a Mixed Population', *Science*, n.s. 28: 49–50

Harris, J. Arthur (1911), 'The Biometric Proof of the Pure Line Theory', *American Naturalist*, 45: 346–63

Herchel, John F. W. (1851), *A Preliminary Discourse on the Study of Natural Philosophy*. London: Printed for Longman, Brown, Green & Longmans

Huxley, Julien (1943), *Evolution: The Modern Synthesis*. New York and London: Harper

Jennings, Herbert S. (1904), *Contributions to the Behavior of Lower Organsims*. Washington, DC: The Carnegie Institute

Johannsen, Wilhelm (1903), *Ueber Erblichkeit in Populationen und in Reinen Linien*. Jena: Gustav Fischer

John Paul II (Pope) (1988), 'Letter to the Reverand George V. Coyne SJ, Director of the Vatican Observatory', *Osservatore Romano* [weekly edition in English], 21(46): 14 November

Kitcher, Philip (2007), *Living with Darwin: Evolution, Design and the Future of Faith*. Oxford: Oxford University Press

Kuhn, Thomas S. (1962), *The Structure of Scientific Revolutions*. Chicago: University of Chicago Press

Laubichler, Manfred, and Jane Maienschein (2007), *From Embryology to Evo Devo: A History of Developmental Evolution*. Cambridge, MA: MIT Press

Luria, S. E., and M. Delbrück (1943), 'Mutations of Bacteria from Virus Sensitivity to Virus Resistance', *Genetics*, 28: 491–511

Lyell, Charles (1830–33), *Principles of Geology: or, The Modern Changes of the Earth and its Inhabitants, Considered as Illustrative of Geology* (3 vols). London: J. Murray

Maienschein, Jane, and Manfred Laubichler (2014), 'Exploring Development and Evolution on the Tangled Bank', in *Evolutionary Biology: Conceptual, Ethical, and Religious Issues*, ed. R. Paul Thompson and Denis M. Walsh. Cambridge: Cambridge University Press

Mayr, Ernst (1942), *Systematics and the Origin of Species*. New York: Columbia University Press

Mendel, Gregor (1865), 'Versuche über Pflanzenhybriden' [Experiments on Plant Hybrids], *Verhandlungen des Naturforschenden Vereins Brünn*, 4

Miescher, Friedrich (1897), *Die histochemischen und physiologischen Arbeiten*. Leipzig: Vogel

Morgan, T. H. (1909), 'For Darwin', *Popular Science Monthly*, 74: 367–80

— (1916), *A Critique of the Theory of Evolution*. Princeton, NJ: Princeton University Press

Muller, H. J. (1918), 'Genetic Variabiltiy, Twin Hybrids and Constant Hybrids in a Case of Balanced Lethal Factors', *Genetics*, 2: 422–99

Nilsson-Ehle, H. (1909), 'Kreuzungsuntersuchungen an Hafer und Weizen', *Lunds Universitets Arsskrift*, n.s., ser. 2, 5(2)

Olby, Robert (1985), *Origins of Mendelism* (2nd edn). Chicago: University of Chicago Press

Pander, Christian (1817), *Dissertatio inauguralis sistens historiam metamorphoseos, quam ovum incubatum prioribus quinque diebus subit*. Wirceburgi: Francisci Ernesti Nitribitt

— (1817), *Beiträge zur Entwickelungsgeschichte des Hühnchens im Eye*. Würzburg: H. L. Brönner

Pearson, K. (1892), *The Grammar of Science*. London: Walter Scott. Two further editions in 1900 and 1911 published by A. & C. Black; Everyman edition published in 1937

— (1900), 'On the Criterion that a Given System of Deviations from the Probable in the Case of Correlated System of Variables is such that it can be Reasonably Supposed to have Arisen from Random Sampling', *Philosophical Magazine*, 50: 157–75

— (1904), 'Mathematical Contributions to the Theory of Evolution, XII: On a Generalised Theory of Alternative Inheritance, with Special Reference to Mendel's Laws', *Philosophical Transactions of the Royal Society*, A, 203: 53–86

Provine, William B. (1971), *The Origins of Theoretical Population Genetics*. Chicago: University of Chicago Press

— (1986), *Sewall Wright and Evolutionary Biology*. Chicago: Chicago University Press

Prusiner, S. B. (1982), 'Novel Proteinaceous Infectious Particles Cause Scrapie', *Science*, 216: 136–44

Queller, David C. (2001), 'W. D. Hamilton and the Evolution of Sociality', *Behavioral Ecology*, 12: 261–4

Richards, Robert J. (1987), *Darwin and the Emergence of Evolutionary Theories of Mind and Behaviour*. Chicago: University of Chicago Press

Ruse, Michael (1979), *The Darwinian Revolution: Science Red in Tooth and Claw*. Chicago: University of Chicago Press

Scott, M. P., and A. J. Weiner (1984), 'Structural Relationships among Genes that Control Development: Sequence Homology between the Antennapedia, Ultrabithorax and Fushi Tarazu Loci of Drosophila', *Proceedings of the National Academy of Science*, 81: 4115–19

Secord, James A. (2000), *Victorian Sensation: The Extraordinary Publication, Reception and Secret Authorship of 'Vestiges of the Natural History of Creation'*. Chicago: Chicago University Press

Simpson, George Gaylord (1951), *The Meaning of Evolution: A Study of the History of Life and of its Significance for Man*. New York: New American Library

— (1951), *Horses: The Story of the Horse Family in the Modern World and through Sixty Million Years of History*. New York: Oxford University Press

— (1980), *Splendid Isolation: The Curious History of South American Mammals*. New Haven, CT: Yale University Press

Spemann, H., and H. Mangold (1924), 'Über Induktion von Embryonalanlagen durch Implantation artfremder Organisatoren', *Wilhelm Roux Archiv für Entwicklungsmechanik der Organismen*. 100: 599–638

Stauffer, R. C., ed. (1975), *Charles Darwin's Natural Selection: Being the Second Part of His Big Book Written from 1856 to 1858*. Cambridge: Cambridge University Press

Stebbins, G. Ledyard (1950), *Variation and Evolution in Plants*. New York: Columbia University Press

Sutton, W. S. (1902), 'On the Morphology of the Chromosome Group in *Brachystola magna*', *Biological Bulletin*, 4: 24–39

— (1903), 'The Chromosomes in Heredity', *Biological Bulletin*, 4: 231–51

Tax, Sol, ed. (1960), *Evolution after Darwin*, vol. I: *The Evolution of Life*. Chicago: University of Chicago Press

— (1960), *Evolution after Darwin*, vol. II: *The Evolution of Man*. Chicago: University of Chicago Press

—, and Charles Callendar, eds (1960), *Evolution after Darwin*, vol. III: *Issues in Evolution*. Chicago: University of Chicago Press

Uglow, Jenny (2002), *The Lunar Men: Five Friends Whose Curiosity Changed the World*. New York: Farrar, Straus & Giroux

Watson, James D. (1968), *The Double Helix: A Personal Account of the Discovery of the Structure of DNA*. New York: Athenium

—, and F.H.C. Crick (1953), 'A Structure for Deoxyribose Nucleic Acid', *Nature*, 171: 737–8

Weinberg, Wilhelm (1908), 'Ueber den Nachweis der Vererbung beim Menschen', *Jahreshefte des Vereins für Vaterländische Naturkunde in Württemburg*, 64: 368–82; English translation in Samuel H. Boyer (1963), *Papers on Human Genetics*, Englewood Cliffs, NJ: Prentice-Hall

Weismann, August (1893), *The Germ-plasm: A Theory of Heredity*, trans. W. Newton Parker and Harriet Rönnfeldt. New York: Scribner

Weldon, W.F.R. (1890), 'The Variations Occurring in Certain Decapod Crustacea, 1: *Crangon vulgaris*', *Proceedings of the Royal Society*, 47: 445–53

— (1894), 'The Study of Animal Variation', *Nature*, 50: 25–6

— (1903), 'Mr Bateson's Revisions of Mendel's Theory of Heredity', *Biometrika*, 2: 286–98

Whewell, William (1847), *Philosophy of the Inductive Sciences*. London: J. W. Parker

Williams, George C. (1966), *Adaptation and Natural Selection: A Critique of Some Current Evolutionary Thought*. Princeton, NJ: Princeton University Press

Wilson, Edward O. (1975), *Sociobiology: The New Synthesis*. Cambridge, MA: Harvard University Press

Wright, Sewall (1931), 'Evolution in Mendelian Populations', *Genetics*, 16: 97–159

Wynne-Edwards, Vero Copner (1962), *Animal Dispersion in Relation to Social Behavior*. New York: Hafner

Yule, G. Udny (1902), 'Mendel's Laws and their Probable Relations to Intra-racial Heredity', *New Phytologist*, 1: 193–207, 222–38

— (1903), 'Professor Johannsen's Experiments in Heredity', *New Phytologist*, 2: 235–42

ACKNOWLEDGEMENTS

My personal interest in evolutionary theory began as an undergraduate. I started my degree in mathematics – a subject I continue to explore and one that has served me well in my career – but my interest soon turned to biology, especially evolutionary biology. Over time, it emerged that it was the conceptual and theoretical richness of evolutionary theory that attracted me. The study of these aspects of biology is mostly undertaken by philosophers. That fact led me to study the broad canvas of philosophy but eventually resulted in a focus on philosophy of science, specifically of biology, which was in its infancy when I entered graduate school. I currently hold appointments in ecology and evolutionary biology, history and philosophy of science, and philosophy; I teach courses in biology and in philosophy of science. The contributions that each area has made to current evolutionary theory will become clear from the narrative.

The watershed in the development of evolutionary theory was the publication in 1859 of Darwin's *On the Origin of Species*. Others had entertained evolutionary views but it was Darwin who provided a theoretical framework that has stood the test of time. He also provided a wealth of evidence. The theory, as he expounded it, was conceptually rich; modern evolutionary theory is vastly richer. Throughout its development evolutionary biology has time and again underscored the brilliance of Darwin and the soundness of the central tenants of his theory. It is hard to overemphasize how impoverished modern biology would be without evolutionary theory as its overarching framework.

Medicine is widely assumed to be a biological enterprise. A sizeable minority of physicians (clinicians and clinical researchers) seem not to accept the fact of evolution or the theory that explains that fact. This I initially found mystifying. About two decades ago, I began doing research in the philosophy of medicine. I now appreciate that clinical medicine is not biological science. It is not theoretically based, which biology, physics and chemistry are. Its reliance on randomized controlled trials has produced at best isolated causal connections between interventions and outcomes. Early on, I thought this to be a function of the complexity of the subject-matter but immunology is every bit as complex and the theoretical richness of that field is impressive. More relevant to this book, evolutionary processes are also deeply complex. Approaching evolutionary biology from a theoretical perspective has yielded a remarkably rich understanding of evolutionary history and the mechanisms that gave rise to that history. So complexity does not explain the impoverished theoretical basis of clinical research. However, it does explain how one can be a brilliant clinician or clinical researcher and yet not accept evolution or appreciate its relevance in clinical medicine.

This volume aims to trace for a non-academic audience the conceptual history of evolutionary theory. Since its audience is non-academic, it does not have the normal academic structure (footnotes, citations and so on). In an academic work, a wealth of footnotes and citations would acknowledge those who have made important contributions to the history and philosophy of evolutionary theory, contributions on which my narrative rests. There are numerous scholarly contributions that have informed the content of this book. My debt to this vast literature is enormous. A few have been so influential on my knowledge and thinking that they warrant acknowledgement here. The work that I have found the most helpful overall is Jean Gayon's *Darwinism's Struggle for Survival* (1998). This is the most comprehensive, detailed, well-documented and philosophically rich source that I have found. William B. Provine's excellent *The Origins of Theoretical Population Genetics* (1971) has enriched chapters Three and Five. Michael Ruse's *The Darwinian Revolution: Science Red in Tooth and Claw* (1979) and Robert Richards's *Darwin and the Emergence of Evolutionary Theories of Mind and Behavior* (1987) provided exemplars for integrating history of science and philosophy of science. I fear that I have not come close to the standard that they set. Ronald Numbers's publications on Darwinism in America enhanced greatly my understanding of the anti-evolution movement in the United States. Jane Maienschein's and Manfred Laubichler's work on development and evolution were invaluable, in important ways, to chapter Ten.

My debts extend well beyond the writings of these and other colleagues. The list of debts is, indeed, too numerous to list. Nonetheless, I think a few are worthy of special mention. My excellent friend of some 50-plus years, Simon Jensen, who is an excellent artist, provided 31 of the illustrations. Martin Young, a professional illustrator, provided the illustration of Darwin's finches. One of my three exceptionally talented children, Kerry Hylton, provided four of the illustrations. My friend from the Netherlands, Gerard Coops, made very helpful comments on an earlier draft. Although he holds a PhD in physics, most of his career has been in consulting. He has an incisive mind. His cheerful, positive attitude is infectious. My colleague Marga Vicedo offered comments on some of the chapters. William (Bill) Mosley encouraged me to use shorter sentences. I have tried to do so, with varying success. He also suggested lead-ins to the chapters, which I have tried to provide. My colleague, Andrew Baines, provided a rich understanding of the role of rhetoric in science. My very good friend Ross Upshur, a clinical physician whose research focuses on public health sciences, has been an excellent sounding board and the source of much valuable inspiration. That he enjoys oysters on the half shell as much I do is an added bonus to our times together. I am also indebted to Ben Hayes, Commissioning Editor at Reaktion Books. First, I am indebted to him for suggesting the book and then patiently waiting for the first, and quite rough, draft. Equally important, he has continually made suggestions that have made the book more readable and accessible to a non-academic audience. Where the book fails in that respect, the blame is all mine. It is not from a shortage of sage advice from him. I enjoyed writing the book much more than I anticipated and I learned an incredible amount along the way.

Finally, a comment I can never say often enough, my spouse of 44 years, Jennifer McShane, is an inspiration and continues to be a wonderful companion on life's journey.

PHOTO ACKNOWLEDGEMENTS

The author and publishers wish to express their thanks to the below sources of illustrative material and/or permission to reproduce it:

Simon Jensen: pp. 22, 32, 33, 35, 39, 70, 71, 74, 86, 110, 114, 133, 134, 142, 145, 150, 155, 158, 159, 161, 162, 166, 167, 172, 173, 176, 178, 187, 188; Kerry Thompson: pp. 30, 53, 109; Martin Young: p. 14.

INDEX